Lanthanide Shift Reagents in Stereochemical Analysis

Methods in Stereochemical Analysis

Volume 5

Series Editor: **Alan P. Marchand**

Department of Chemistry
North Texas State University
Denton, Texas 76203

Distribution: VCH Verlagsgesellschaft mbH, P.O. Box 1260/1280, D-6940 Weinheim,
Federal Republic of Germany

USA and Canada: VCH Publishers, Inc., 303 N.W. 12th Avenue, Deerfield Beach, FL 33442-1705, USA

Lanthanide Shift Reagents in Stereochemical Analysis

Edited by
Terence C. Morrill

Terence C. Morrill
Rochester Institute of Technology
Rochester, New York 14623

Library of Congress Cataloging-in-Publication Data

Lanthanide shift reagents in stereochemical analysis.

(Methods in stereochemical analysis; v. 5)
Includes bibliographies and index.
1. Chemical tests and reagents. 2. Nuclear magnetic resonance spectroscopy. 3. Rare earth metals. 4. Stereochemistry. I. Morrill, Terence C. II. Title. III. Series.
QD77.L36 1986 543'.0877 86-7752
ISBN 0-89573-119-3

Printed in the United States of America.

0-89573-119-3 VCH Publishers
3-527-26167-2 VCH Verlagsgesellschaft

Methods in Stereochemical Analysis
Volumes in the Series

1 Stereochemical Applications of NMR Studies in Rigid Bicyclic Systems
by *Alan P. Marchand*

2 Carbon-Carbon and Carbon-Proton NMR Couplings: Applications to Organic Stereochemistry and Conformational Analysis
by *James L. Marshall*

3 Stereochemistry and Reactivity of Systems Containing π Electrons
by *William H. Watson*

4 Applications of Dynamic NMR Spectroscopy to Organic Chemistry
by *Michinori Oki*

5 Lanthanide Shift Reagents in Stereochemical Analysis
Edited by *Terence C. Morrill*

6 Applications of NMR Spectroscopy to Problems in Stereochemistry and Conformational Analysis
Edited by *Yoshito Takeuchi* and *Alan P. Marchand*

7 High Resolution NMR Spectroscopy of Synthetic Polymers in Bulk
Edited by *Richard A. Komoroski*

8 Phosphorus-31 NMR Spectroscopy in Stereochemical Analysis: Organic Compounds and Metal Complexes
Edited by *John G. Verkade* and *Louis D. Quin*

9 Two-Dimensional NMR Spectroscopy
Edited by *William Croasmun* and *Robert Carlson*

Contributors

Chapter 1:
Terence C. Morrill
Department of Chemistry,
Rochester Institute of Technology
One Lomb Memorial Drive
Rochester, New York 14623

Chapter 2:
Julien B. Chiasson
and
Krzysztov Jankowski
Département de Chimie
Université de Moncton
Moncton, New Brunswick E1A 3E9
Canada

Chapter 3:
Douglas J. Raber
Department of Chemistry
University of South Florida
Tampa, Florida 33620

Chapter 4:
Jaakko Paasivirta
Department of Chemistry
University of Jyväskylä
Seminaarinkatu 15
SP-40100
Jyvaskyla 10
Finland

Chapter 5:
Thomas J. Wenzel
Department of Chemistry
Bates College
Lewiston, Maine 04240

Preface

The use of lanthanide shift reagents (LSRs) as applied to chemical analysis using nuclear magnetic resonance is not an extremely new topic. Organic chemists have been developing this type of analysis for a decade and a half. Although spectroscopists were aware earlier of the NMR effects of paramagnetic species, it was clearly Hinckley's 1969 publication (ref. 1, Ch. 1) that caused the explosion of interest in the field. The 1970s saw an extensive development of the field, and standard textbooks and lab manuals now incorporate LSR analysis as part of their fare.

It now seems appropriate to review the use of LSR compounds in the context of stereochemical analysis. The detailed analysis of the configuration of organic compounds represents both an ever-present area for LSR application and a proving ground for refinement of LSR theory.

We have prepared in order chapters which present the following kinds of information. First, I have written an introductory chapter. My purpose was threefold: I felt it was important to outline LSR compounds and theory for the uninitiated. This was a quite straightforward task as my own interests have kept me aware of the many good reviews (Ch. 1, refs. 3–19) in the field. I also tried to paint a backdrop for the subsequent chapters and finally to provide references for more specialized approaches.

In the second chapter Krzysztoff Jankowski and Julien B. Chiasson discuss their computer-aided approaches to determining the geometry of LSR–substrate complexes. In the earliest LSR studies it became clear that both angular and distance factors influenced the magnitude of LSR-induced effects. In fact some significant success has been obtained in analyzing ^{1}H NMR applications where largely pseudocontact interactions can be assumed. Jankowski and Chiasson guide you through the setup of such analyses and then discuss typical results obtained for most standard organic functional groups. Computer results are displayed for stereochemical analysis of specific molecules and the validity of these results is examined. A line by line listing of their program (in APL, including subroutines) is provided. A discussion of six other computer approaches is also included.

In the third chapter, Douglas Raber takes another approach to scrutinizing the nature of the LSR–substrate complex. He has attacked the very challenging problem of the involved equilibria arising from the various complexes between substrates and LSR compounds. Building on the fact that more than one stoichiometry can exist between an LSR and a substrate, he describes how to sort out these complexes, including the crucial differentiation between observed shifts and bound shifts. This, by definition, must be and is a chapter that scrutinizes the usually very rapid exchange between LSR–substrate complexes and their free components and examines the very few cases in which these exchanges have been slowed on the NMR time scale. A step by step discussion of the four analytical methods that can be used to determine bound shifts follows, describing the fundamental assumptions behind these methods; from this information recommendations can be made of where these methods may apply. This is followed by experimental results on stereochemically well-defined systems, and the subsequent error analysis allows for conclusions and recommendations to be made regarding these methods. Some discussion of ^{13}C NMR analysis, and ^{1}H NMR analysis on polyfunctional compounds concludes this chapter, a chapter of very wide scope.

Chapter 4 includes the kind of work for which its author, Jaako Paasivirta, is well known. He describes in thorough detail the solution of complex stereochemical problems using LSR compounds. Such solutions include discussions of experiments, choice of the stereochemical problems, the detailed synthesis and other chemistry behind the substrates of choice, the computer program (LANTA) he has used, and detailed results for certain bridged polycyclic compounds. Stereochemical analysis of diastereotopic groups, ring conformation, deuterium labeling, and other labeling is then outlined. This chapter thus thoroughly examines the area that aligns most closely with the title of the book itself.

In the fifth chapter, Thomas Wenzel describes efforts to carry out LSR analysis on substrates that have eluded analysis until 1975. Specifically, most earlier LSR analysis had been limited to reasonably basic organic substrates, such as amines, alcohols, ethers, and carbonyl-containing compounds. This is because LSR compounds with lanthanide cations of 3+ charge are hard acids (on the Pearson HSAB scale) and so coordinate more strongly with hard bases or, among soft organic base substrates, those of reasonably high basicity. Therefore, entire classes, of organic compounds (thioethers, mercaptans, nitro compounds, alkenes, aromatic hydrocarbons) resisted analysis; these compounds contained highly polarizable Lewis base sites and/or were inherently weakly basic and so did not interact strongly with LSR compounds. Wenzel (and his doctoral mentor R. E. Sievers) have led the way in showing that involvement of the softer Lewis acid, Ag(I), greatly enhances LSR analysis. Therefore combinations of Ag(I)–Ln(III) diketonate complexes have allowed analysis of the uncooperative substrates listed above. This reagent class has been so successful, it is presently commercially available (Aldrich).

It is hoped that we thus present an up to date account of progress in an extremely active area.

Terence C. Morrill

Contents

Preface **ix**

Chapter 1: An Introduction to Lanthanide Shift Reagents
Terence C. Morrill **1**

Chapter 2: Computer Approaches to the Geometry of the LSR-Substrate Complex: Geometry of Shift Reagents-Substrate Complexes, Simulation of Induced Chemical Shifts
Julien B. Chiasson and Krzysztof Jankowski **19**

Chapter 3: The Nature of the LSR-Substrate Complex
Douglas J. Raber **55**

Chapter 4: Solutions to Stereochemical Problems
Jaakko Paasivirta **107**

Chapter 5: Binuclear Lanthanide(III)-Silver(I) NMR Shift Reagents
Thomas J. Wenzel **151**

Author Index **175**

Subject Index **185**

1

AN INTRODUCTION TO LANTHANIDE SHIFT REAGENTS

Terence C. Morrill

Department of Chemistry,
Rochester Institute of Technology,
Rochester, New York 14623

The field of lanthanide shift reagents (LSRs) has been reviewed extensively and this chapter is not an attempt to reproduce those reviews. Instead, this represents an effort to introduce LSRs to the uninitiated, to present a lead into the subsequent chapters, and finally to provide references to the literature.

By far the most common LSRs are those abbreviated $Eu(dpm)_3$, **1**, and $Eu(fod)_3$, **2** (where dpm is the *tris*-dipivaloylmethanato derivative and fod is the *tris*-6,6,7,7,8,8,8-heptafluoro-2,2-dimethyloctanetonato derivative). Although the earliest reference[1] in the field described the use of a dipyridine complex, $Eu(dpm)_3(py)_2$, where py is pyridine, later it became clear that **1** and **2** were much more useful.[2] A comprehensive understanding of the theory and application of these reagents may be obtained from the many reviews in this area.[3–19]

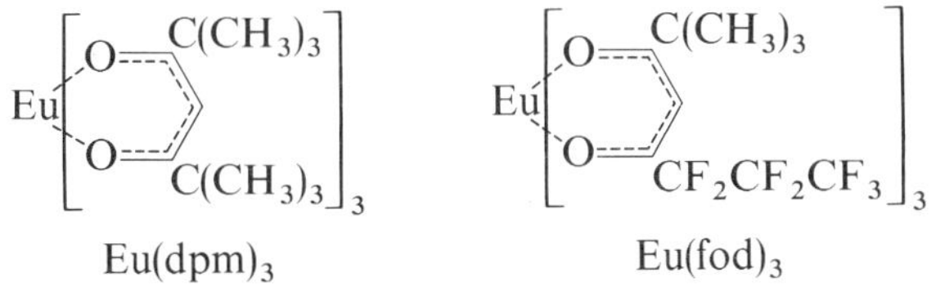

Eu[O…C(CH3)3 / O…C(CH3)3]3 — $Eu(dpm)_3$

1

Eu[O…C(CH3)3 / O…CF2CF2CF3]3 — $Eu(fod)_3$

2

Morrill (ed): LANTHANIDE SHIFT REAGENTS IN STEREOCHEMICAL ANALYSIS

It is important to recognize certain characteristics of the LSR structure that make these complexes useful. It was known that certain paramagnetic ions, such as Eu^{3+} and Pr^{3+}, would induce shifts in ^{1}H and other NMR signals.[20] However, organic applications became commonplace only after complexes with organic ligands, normally of the acetonylacetonate (or acac) type, as in **1** and **2**, were prepared. Acetonylacetonate-type ligands, such as dpm and fod, cause the LSR reagents to be soluble in NMR solvents, such $CDCl_3$ and CCl_4.

Among the triply charged lanthanide cations only La^{3+} and Lu^{3+}, with respectively $4f^0$ and $4f^{14}$ shells, are diamagnetic. Among the paramagnetic ions Pr^{3+} ($4f^2$), Eu^{3+} ($4f^6$) and Yb^{3+} ($4f^{13}$) have short electron-spin relaxation times ($<10^{-12}$ s) and so induce NMR shifts without inducing appreciable line broadening. Alternatively, Gd^{3+} ($4f^7$) and Eu^{2+} ($4f^7$), with long electron-spin relaxation times ($>10^{-10}$ s), are used to enhance NMR relaxation rates (without appreciable shifts). The latter two cations, therefore, although of no value for lanthanide induced shift (LIS) studies, have been used to induce quantitative signals in ^{13}C NMR analysis. The ions Dy^{3+} ($4f^9$) and Ho^{3+} ($4f^{10}$), with intermediate electron-spin relaxation times, also enhance relaxation rates and yet can still be used to induce shifts.

The magnetic anisotropy of the LSR is felt by protons of organic compounds because the Lewis-acidic metal cations coordinate with Lewis-basic sites of functional groups in these compounds, for example:

$$\begin{array}{c} \delta- \\ R-\ddot{O}\text{:----}Eu^{3+} \\ | \\ H \end{array}$$

It is well known that there are at least two sites available in such complexes as **1** and **2** that can be occupied by alcohols or other organic substrates the NMR signals of which are to be shifted.

A common and simple way to carry out an LSR analysis is to add measured increments of **1** or **2** to a known amount of an organic compound dissolved in either $CDCl_3$ or CCl_4. In this way Sanders and Williams[21] were able to convert even the overlapping ^{1}H NMR signals (in the absence of LSR) to a spectrum in which a well-separated first-order signal could be related to each of the different sets of C–H protons of a 1-alkanol (Figure 1-1). This occurs because the LSR enhances the nonequivalence of the protons in the alcohol while causing virtually no change in the interset coupling constants. In effect, $\Delta v/J$, the ratio of the frequency difference to the coupling constant for these protons, is greatly increased. Typically, mole ratios of LSR to organic substrate are in-

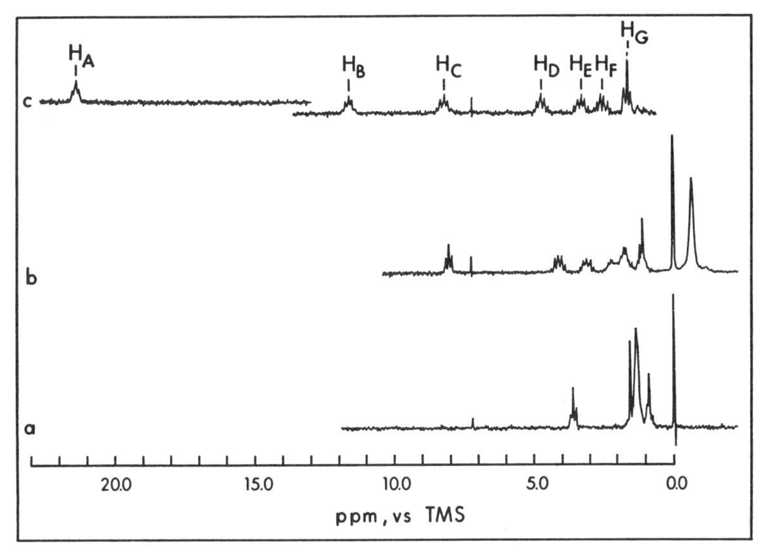

Figure 1-1. Proton NMR spectrum (60 MHz) of 1-heptanol (a) 0.3 M alcohol in $CDCl_3$; (b) same as (a) with 0.78 M $Eu(dpm)_3$.

G F E D C B A
$CH_3CH_2CH_2CH_2CH_2CH_2CH_2OH$

(From Rabenstein, D. L. *Anal. Chem.*, **1971**, *43*, 1599.)

creased over a range of 0.1–0.8. Increases above a 1:1 ratio frequently lead to precipitate formation.

A simple approach to this coordination process involves the following equilibrium. In this process

$$LSR + S \rightleftharpoons LSR{\cdot}S \qquad (1\text{-}1)$$

the Lewis-acidic lanthanide of the LSR coordinates with a Lewis-basic site in the substrate, S. This coordination is a rapid reversible process and so only in rare cases can separate signals be seen for both S alone and for S in LSR·S.[22] Instead, normally a weighted average of contributions from S and LSR to a single signal is observed for each set of protons in the substrate. As larger amounts of LSR are added, greater contributions from LSR·S are realized, leading to larger LISs. Clearly, the NMR solvent must be much less Lewis basic than the substrate for the LSR to coordinate primarily with S.

It is clear that the position of the equilibrium (eq. 1-1) depends on the basicity of substrate S. A feeling for this may be obtained from Table 1-1. In that table, as expected, the more basic substrates show larger LIS magnitudes per mole of added LSR. An aspect of this basicity includes the Pearson hard–soft acid–base concept (HSAB).[23] That is,

TABLE 1-1. TYPICAL LIS VALUES FOR SUBSTRATES IN CCl_4 SOLUTION[11]

Functionality	G[a](ppm)	Functionality	G[a](ppm)
$RCH_2N\underline{H}_2$	150	$RCOOC\underline{H}_3$	14
$RCH_2O\underline{H}$	100	$RC\underline{H}_2COOCH_3$	13
$RC\underline{H}_2NH_2$	35	$(C\underline{H}_3)_2SO$	8.9
$RC\underline{H}_2OH$	25	$(RC\underline{H}_2O)_3PO$	8.8
$(CH_3)_2N\cdot CO\cdot \underline{H}$	34	$(RO)_2PO\cdot C\underline{H}_3$	7
$(CH_3)_2N\cdot CO\cdot CH_3$	12.5	$RC\underline{H}_2CN$	5
$RCO\cdot \underline{H}$	19	$(RC\underline{H}_2)_2S$	<1
$(RC\underline{H}_2)_2CO$	11	$RC\underline{H}_2OX^b$	<1
$(RC\underline{H}_2)_2O$	10		

[a] LIS Gradient: ppm per mol LSR per mol subst rate.
[b] $X = CPh_3$, $COCF_3$ or *p*-toluenesulfonyl.

the LSR possessing an Eu^{3+}, a hard–acid site, coordinates more strongly to and shows greater LIS values for harder (for example, oxygen–ether) bases than softer (for example, thio-ether) bases.[24]

It must be kept in mind that such data as those in Table 1-1 are the results of very complex solution equilibria. That is, in addition to the LSR binding constants influencing the equilibrium position in eq. (1-1), a number of other equilibria must be considered, including further complexation of the substrate with the LSR (eq. 1-2) and LSR to LSR complexation (eq. 1-3).

$$LSR\cdot S + S \rightleftharpoons LSR\cdot S_2 \qquad (1\text{-}2)$$

$$LSR + LSR \rightleftharpoons (LSR)_2 \qquad (1\text{-}3)$$

It also should be kept in mind that a low concentration of LSR impurities, such as water, can compete strongly with the substrate for the Ln cation site in the LSR. For example, plots of LIS ($\Delta\delta$) vs. [LSR]/[S] under 0.1–0.2 become nonlinear because of water or other impurities and become nonlinear at [LSR]/[S] above 0.5–0.7 because of other competing equilibria (Figure 1-2). Finally, some of the LIS data that have been reported, especially in earlier publications, should be viewed with caution because the LSR purity may be in question.

The fact that early researchers chose europium complexes was fortunate because added Eu–LSRs cause a minimum number of crossovers in LIS studies. With other (eg, Pr^{3+}) shift reagents added increments may very likely cause one signal to overlap with rather than separate from another signal. Specifically, as typified with 1-alkanols, in the absence of shift reagents the electronegative hydroxy substituent causes the adjacent protons to be more downfield than successively more re-

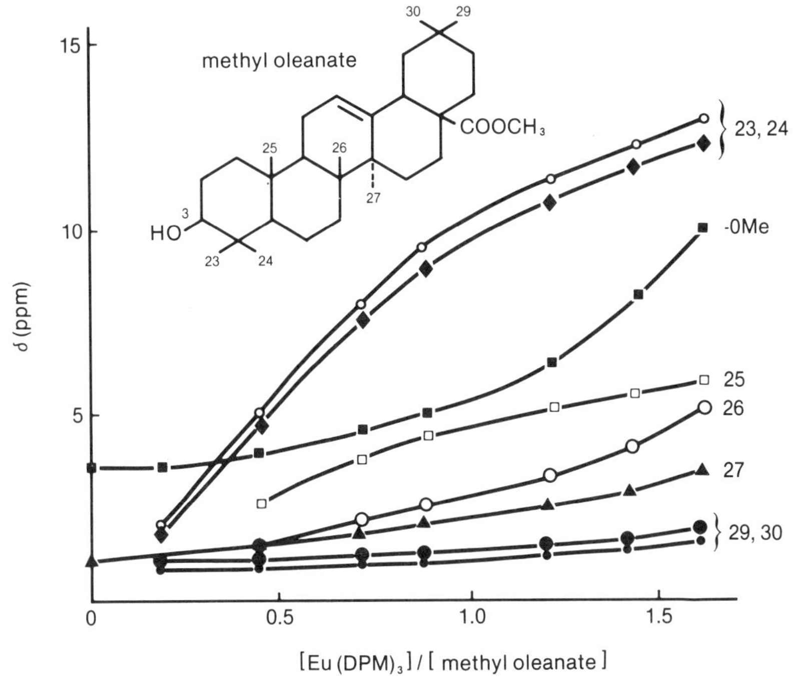

Figure 1-2. The effect of added $Eu(dpm)_3$ on the methyl resonances of methyl oleanate (approximately 0.09 M in CCl_4 solution.) (From ref. 29.)

mote protons in the structure. As Eu LSR is added, downfield protons are shifted even more downfield and normally are shifted most when they are closest to the Lewis basic (for alcohols, hydroxy) site. The various signals therefore frequently show enhanced separations. It is clear, however, that the distance between the proton and the LSR is not the only factor that influences the LIS and so it is sometimes useful to employ praseodymium, which usually induces upfield shifts.

A number of operational points regarding shift reagent experiments should be discussed. The NMR solvent should not compete significantly with the substrate; therefore one normally is constrained to using $CDCl_3$, CCl_4, and C_6D_6. In the absence of added substrate $Eu(dpm)_3$ has a solubility of 40 mg/mL in $CDCl_3$, whereas added alcohol substrate increases solubility to 200–300 mg/mL in $CDCl_3$ and C_6D_6 and to ca. 100 mg/mL in CCl_4.[2,25] In contrast, $Eu(fod)_3$ has a solubility of ca. 400 mg/mL.[25,26] The solubility of the LSR–S complex may be different from, and can be higher than, the LSR alone. Typical proton resonances of the LSR in the presence of substrate are: $Eu(fod)_3$, δ 0.4–2.0; $Eu(dpm)_3$, δ – 1.0-2.0; $Pr(dpm)_3$, δ 3.0–5.0. In $Eu(dpm)_3$–(alcohol) complexes, the *t*-butyl group occurs at $\delta - 0.70$.[25] Deuterated reagents that avoid these interferences are available commercially. Because the lanthanide cations have extremely high magnetic susceptibilities, use of an internal standard is an absolute requirement.[5] Sievers and coworkers[27,28] have described the preparations of $Ln(dpm)_3$- and $Ln(fod)_3$-type reagents. All of these reagents should be stored in desicators and/or sublimed before use.[29,30]

The vast majority of proton NMR shifts can be described by the sum of two terms arising from contact (C) and pseudocontact (PC) effects:

$$\Delta\delta = \Delta\delta_C + \Delta\delta_{PC} \tag{1-4}$$

The former effect normally is not important for proton shifts. Carbon shifts should be predicted using a complex formation term ($\Delta\delta_{CF}$) as well.

The majority of ^1H NMR shifts can be rationalized using a modified form of the McConnell-Robertson equation:

$$\Delta\delta_{PC} = \frac{K(3\cos^2\theta - 1)}{r^3} \tag{1-5}$$

This equation is valid only if the complex displays apparent axial (or at least threefold) symmetry; that is, the magnetic susceptibility factors in both the x and y directions should be equal.[6,7] In the modified equation, $\Delta\delta_{PC}$ is the pseudocontact LIS, K is a constant characteristic of the LSR·S complex, r is the length of a paramagnetic vector between the center of the nucleus (with whose shift we are concerned) and the paramagnetic lanthanide ion, and θ is the angle between vector r and the principle (or z) magnetic axis of the complex. It has been found useful to assume that the principal magnetic axis is virtually colinear with the axis through the center of the lanthanide ion and the substrate atom to which the lanthanide coordinates (in Figure 1-3a, oxygen is the coordinating substrate atom). This approach has been supported by experimental investigation.[31–33]

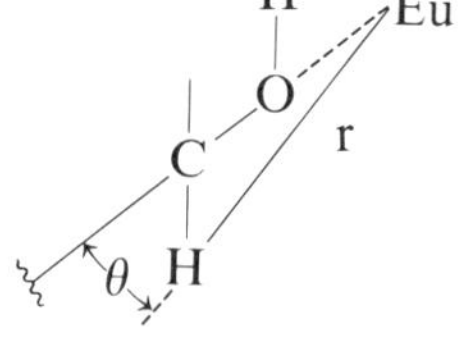

H Eu O r C θ H

Figure 1-3a

In the solid state, $Eu(dpm)_3(py)_2$ has been shown not to have threefold symmetry as determined by x-ray analysis.[34] In solution, apparently, rapidly interconverting geometries can average out to an "effective" axial symmetry.[35]

A test for the pseudocontact mechanism for a given LSR·S complex may be carried out by comparing the LIS ratios for two different nuclei in the same complex to the same ratio using a different LSR. The ratios are expressed as follows.

$$\frac{[(\text{LIS})_{\text{LSR}_1}]_i}{[(\text{LIS})_{\text{LSR}_1}]_j} = \frac{K_{\text{LSR}\cdot\text{S}}\,(3\cos^2\theta_i - 1)/r_i^3}{K_{\text{LSR}\cdot\text{S}}\,(3\cos^2\theta_j - 1)/r_j^3} = \frac{[(3\cos^2\theta - 1)/r^3]_i}{[(3\cos^2\theta - 1)/r^3]_j}$$

and

$$\frac{[(\text{LIS})_{\text{LSR}_1}]_i}{[(\text{LIS})_{\text{LSR}_1}]_j} \Bigg/ \frac{[(\text{LIS})_{\text{LSR}_2}]_i}{[(\text{LIS})_{\text{LSR}_2}]_j} = \frac{[(3\cos^2\theta_1 - 1)/r_1^3]_i}{[(3\cos^2\theta_1 - 1)/r_1^3]_j} \Bigg/ \frac{[(3\cos^2\theta_2 - 1)/r_2^3]_i}{[(3\cos^2\theta_2 - 1)/r_2^3]_j}$$

Therefore we see that the K values cancel and the LIS ratio depends only on a constant ratio of geometric values that is independent of the LSR employed.

Complex formation shifts, important for ^{13}C-NMR in addition to the Δ_{PC} and Δ_C terms mentioned above, result from structural changes in the substrate molecule induced by complexation with the lanthanide ion. This complexation results in polarization of electron clouds in the substrate and affects the electronic state of the substrate nuclei, inducing shifts in their NMR frequencies.

Equation (1-5) successfully describes LISs that are the result of a pseudocontact mechanism; the vast majority of ^{1}H LIS shifts follow this equation. Carbon-13 nuclei give rise to LIS values that contain substantial contributions from contact shifts, whereas contact contributions to ^{1}H LIS values normally are negligible.

The contact shift involves delocalization of the free electron density from the paramagnetic lanthanide into the organic substrate; eq. (1-6) describes this:

$$\Delta\delta_C = \frac{-2\pi\beta\nu AJ(J+1)g_L(g_L - 1)}{3kT\gamma} \tag{1-6}$$

where ν and γ are, respectively, the Larmor frequency and the magnetogyric ratios of the nucleus of interest; β is the Bohr magneton; J is the electronic spin angular momentum; g_L is the Lande g value; and A is the scalar coupling constant. The contact interaction takes place only when there is a measurable probability of finding unpaired electron spin density on the atomic s orbital of the nucleus of interest.

For protons the contact contributions are much smaller than for ^{13}C because the scalar coupling constant is smaller, among other factors.[5]

The mechanism of the contact interaction involves direct overlap between the metal orbital bearing the free electron density and the orbital of the coordinated substrate. There is a sizeable contact contribution from overlap between the *d* orbitals of transition-metal ions and organic substrates. In lanthanide cations such as Eu^{3+}, however, *f* orbitals bear the free-electron density and these orbitals are highly shielded. Therefore, coordinating substrates form weak covalent bonds, and hyperfine couplings with such lanthanide cations are about an order of magnitude less than with transition-metal cations. This is also related to the electron-spin relaxation times induced by lanthanide cations, as discussed above.

Very early interpretations of the pseudocontact shift in proton NMR spectra overemphasized the distance factor in eq. (1-5). For example, plots of log $\Delta\delta$ vs. log r (based on eq. 1-5) were examined and nearness to a slope of 3 was taken as a measure of the importance of the pseudocontact interaction. This neglects the importance of the angular term. Early workers probably were mislead by the apparent success of this approach caused by linear geometries for an LSR·S complex, such as in Figure 1-3b, in which the difference between the angles θ and θ' was too small to make a significant contribution to the LIS. Shapiro et al.,[36] however, showed how omission of the angle dependence can lead to serious errors.

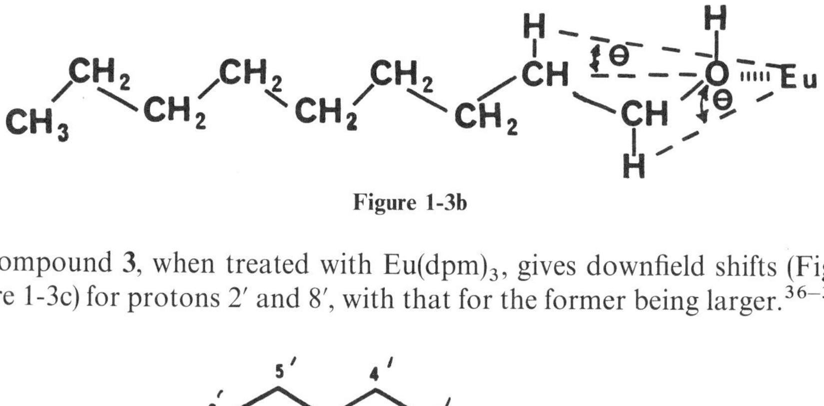

Figure 1-3b

Compound **3**, when treated with $Eu(dpm)_3$, gives downfield shifts (Figure 1-3c) for protons 2′ and 8′, with that for the former being larger.[36–38]

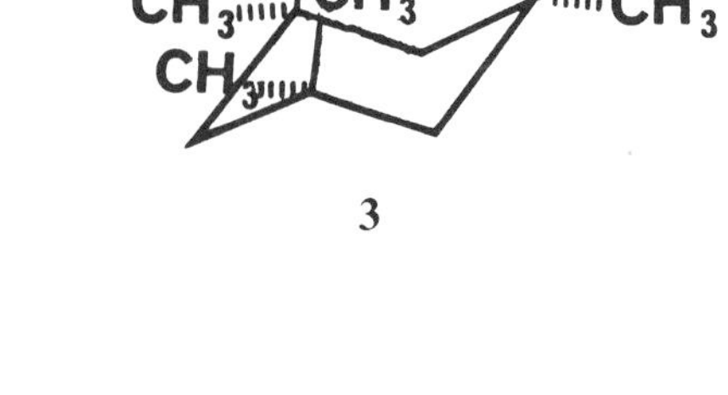

3

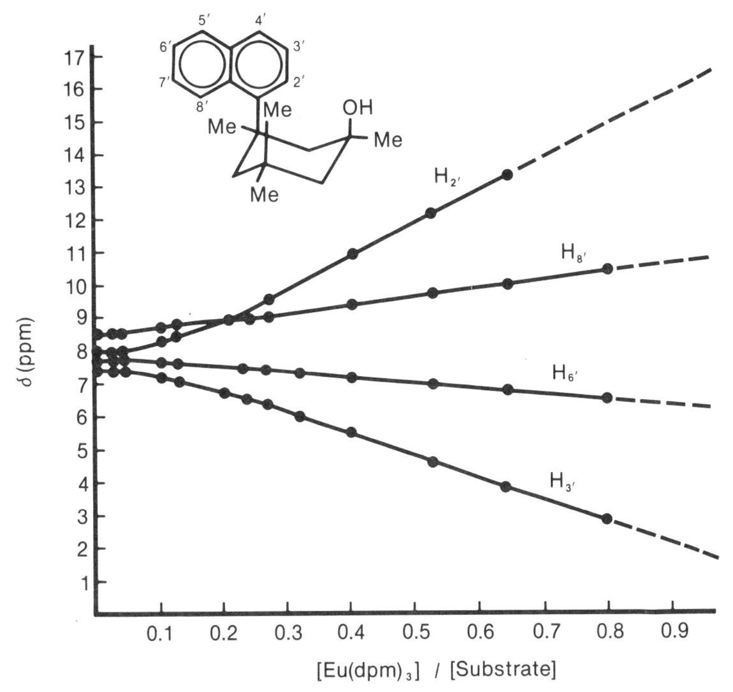

Figure 1-3c. Proton NMR shifts of the naphthyl protons of *cis*-1,3,5,5-tetramethyl-5-(1-naphthyl)cyclohexan-1-ol as a function of the $Eu(dpm)_3$ concentration. (From ref. 36.)

It is tempting simply to conclude that this is because the $H_{2'}$ is closer to the hydroxyl group and so closer to the coordinated Eu atom. It is clear, however, that the angle dependence must be considered, because $H_{4'}$ and $H_{3'}$ show upfield LIS values in the same experiments. These should be interpreted with the aid of Figure 1-4. If the angle (θ) between r and the principal magnetic axis of the complex is between 54.7° and 125.3°, the sign (and thus the direction) of $\Delta\delta$ from eq. (1-5) is opposite to that found for other values of θ. Also, near 54.7° and 125.3° the

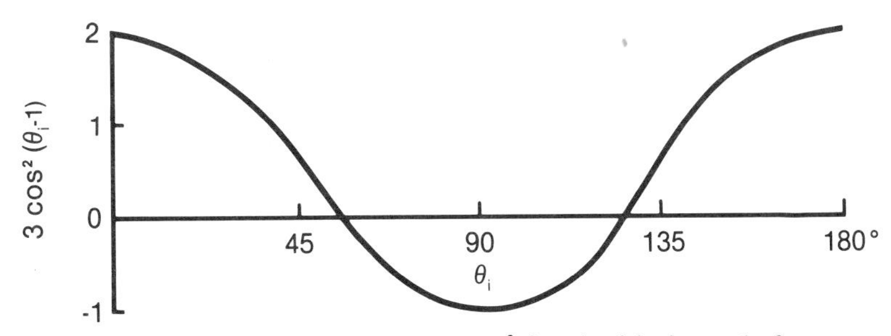

Figure 1-4. The variation of $3 \cos^2 \theta_i - 1$ with the angle θ.

magnitude of the shift is smaller. Compound **3** therefore apparently has angles for $H_{4'}$ and $H_{8'}$ in the 54.7–125.3° range.

Structural characteristics of the shift reagent *vis-à-vis* the nature of the ligands and the lanthanide atom control the chemical characteristics of the reagent both in the solid state and in solution, as well as in the presence or absence of the substrate.

The relative importance of the lanthanide atom in inducing the LIS can be seen in Table 1-2. When cyclohexanone is used as a substrate it is apparent that $Pr(dpm)_3$ is somewhat more effective at inducing upfield shifts than $Eu(dpm)_3$ is for downfield shifts. It also may appear from Table 1-2 that $Tb(dpm)_3$ or $Ho(dpm)_3$ are superior reagents (Table 1-2); however, from scrutiny of Table 1-3, it can be seen that Tb^{3+} and Ho^{3+} both induce substantial line broadening. It is clear here that line broadening also is induced by transition-metal ions.

Table 1-4 displays the proportion of contact vs. pseudocontact shift caused by various lanthanide cations. These effects are especially important with ^{13}C, ^{14}N, ^{15}N, ^{17}O, ^{19}F, and ^{31}P nuclei that are close to the point of coordination with the lanthanide [^{13}C, ^{19}F;[7] ^{31}P;[18] ^{14}N].[39] It also seems that nuclei of more highly polarizable substrates (such as pyridine) are more susceptible to contact shifts.[40] Contact shifts can be detected by observing an alternation of the sign of the LIS for a sequence of a given type of nucleus in a substrate. For example, carbons successively more remote from the functional group (Table 1-5) show this sign alteration: $Eu(fod)_3$ induces a shift for C-2 of 2-aminobutane in one

TABLE 1-2. LIS Values of Shifts Induced by Various Lanthanides on the α- and γ-Methylene Groups of Cyclohexanone[3]

	Shift caused by $M(dpm)_3$	
	α-Methylene (Hz)	γ-Methylene (Hz)
Pr	−11.25	−3.7
Nd	−5.55	−1.8
Sm	−1.35	−0.6
Eu	2.95	1.8
Gd	—	—
Tb	−26.25	−10.9
Dy	−54.00	−17.9
Ho	−51.45	−18.1
Er	25.55	8.8
Tm	−44.65	−14.8
Yb	12.15	4.4
Lu	0.00	0.0

TABLE 1-3. MAGNITUDE OF NMR LINE BROADENING INDUCED BY LANTHANIDE AND TRANSITION METAL IONS[3]

Lanthanide	Bandwidths (Hz)[a]	Half-height bandwidths (Hz)[b]	Relative broadening (Hz per Hz of shift)[c]
Pr	40	5.6	0.005
Nd	16	4.0	—
Sm	7	4.4	0.02
Eu	10	5.0	0.003
Gd	1500	—	—
Tb	250	96.00	0.1
Dy	180	200.00	—
Ho	180	50.00	0.02
Er	250	50.00	—
Tm	400	65.00	—
Yb	60	12.00	0.02
Transition metal as M(III)	**Bandwidth (Hz)[d]**		
Ti	2000		
V	25		
Cr	1000		
Mn	100		
Fe	800		
Mo	200		
Ru	100		

[a] Lanthanide bandwidths of *t*-butyl in $M(dpm)_3$ in carbon tetrachloride.
[b] Lanthanide half bandwidths of methyl in 2-picoline using $M(dpm)_3$.
[c] Lanthanide relative broadening of *t*-butyl in $M(dpm)_3$.
[d] Transition-metal bandwidths of methyl in $M(acac)_3$.

TABLE 1-4. RATIO OF CONTACT AND PSEUDOCONTACT SHIFTS[19, a]

Ln^{3+}	Theory	Obs	Obs
Pr	0.10	0.34	0.31
Nd	0.40	0.91	0.61
Sm	−0.03		−0.26
Eu	1	1	1
Tb	−0.14	0.17	−0.26
Dy	−0.11	0.13	−0.13
Ho	−0.22	0.09	−0.25
Er	0.17	0.30	0.27
Tm	0.06		0.06
Yb	0.04	0.09	0.06

[a] Scaled to unity for Eu^{3+}.

TABLE 1-5. LANTHANIDE-INDUCED ^{13}C SHIFTS (IN ppm) FOR 2-AMINOBUTANE[11]

	C_1	C_2	C_3	C_4
$Eu(dpm)_3$	−24.2	−90.2	+11.0	−14.4
$Eu(fod)_3$	−12.7	−81.9	+23.6	−8.9
$Pr(dpm)_3$	+54.5	+121.0	+34.6	+23.9
$Pr(fod)_3$	+49.9	+111.0	+27.7	+23.6
$La(dpm)_3$	+2.1	~0	−0.6	+0.2

direction and for C-3 a shift in the opposite direction. Calculations using the INDO MO method support this conclusion. Magnetically active nuclei with nonbonded pairs of electrons are expected to experience substantial contact shifts. Returning to changes in the lanthanide ion, it is clear from Table 1-4 that Yb reagents should show LIS caused largely to the pseudocontact mechanism. This, coupled with the fact that the magnitude of observed shift is substantial (Table 1-2), makes $Yb(dpm)_3$ reagents desirable for studies of the geometry of LSR complexes.

As mentioned earlier, "complex formation" shifts can contribute to the magnitude of the LIS along with shifts caused by pseudocontact and contact interactions. The reagent $La(dpm)_3$ normally is used to measure complex formation shifts.[39] It has been found that these shifts are negligible for ^{1}H, but the ^{13}C nuclei closest to the site of coordination of lanthanide coordination show measurable shifts for both alcohols (3–5 ppm per mole of LSR per mole S) and ketones (7–12 ppm), and all are downfield shifts.[11,41]

An important use for changes in the lanthanide atom is a check on the applicability of the pseudocontact mechanism. It has been shown[39] that several of the protons and carbons of cholesterol give a constant ratio of LIS values for a given pair of LSR reagents. Specifically, the $Eu(dpm)_3$–$Dy(dpm)_3$ ratio was used, and assuming a pseudocontact shift mechanism for each and the same geometry (θ, r) for the two LSR·S complexes:

$$\frac{\text{LIS}_{Eu}}{\text{LIS}_{Dy}} = \frac{K_{Eu}\,(3\cos^2\theta - 1)/r^3}{K_{Dy}\,(3\cos^2\theta - 1)/r^3} = \frac{K_{Eu}}{K_{Dy}}$$

The ratio therefore should be a constant for each nucleus.[42,43] An independent check of the distance (r) portion of the geometry factor can be made using Gd^{3+}.[33]

If additional ligands block the LSR Lewis acid site in the LSR, the LSR becomes less effective. Sanders and Williams[21] showed that $Eu(dpm)_3$ was four times more effective than $Eu(dpm)_3(py)_2$ in coordinating with cholesterol. Other comparable ligands that have been used include dibenzoylmethanato ($C_6H_5COCHCOC_6H_5$)[2,44] and benzoylacetonato ($C_6H_5COCHCOCH_3$)[41] with no apparent advantage over dpm. Moreover it has been proposed that the bulky *t*-butyl groups in the dpm ligands restrict orientation in complex formation, preventing the magnetic susceptibility tensors from being averaged out and allowing measurable pseudocontact shifts.[45]

The Lewis acidity of the LSR is dependent sensitively on substituents on the ligands of the LSR. As mentioned earlier fod usually causes the LSR to complex more strongly than dpm. The isomers of *p*-methoxy-*p*′-butylazoxybenzene, **4** and **5**, display coincidental overlap of the methoxy hydrogens.

$CH_3CH_2CH_2CH_2C_6H_4N{=}N(\rightarrow O)C_6H_4OCH_3$ **4**

$CH_3CH_2CH_2CH_2C_6H_4N(\rightarrow O){=}NC_6H_4OCH_3$ **5**

Addition of $Eu(dpm)_3$ does not resolve these signals. However, addition of only 10 mg of $Eu(fod)_3$ to a 0.2 M solution of these isomers induces LIS values sufficient for resolution. A similar result is obtained with *p*-methoxy-*p*′-methyl compounds (Figure 1-5).

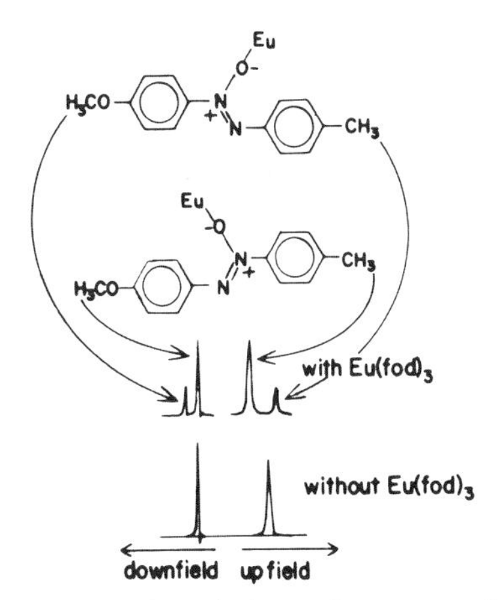

Figure 1-5. Isomers of *p*-methoxy-*p*′-methylazoxybenzene and partial NMR spectra of the isomeric mixture showing upfield and downfield shifts induced by the addition of $Eu(fod)_3$. (From ref. 15.)

To many chemists it has seemed attractive to use even more highly fluorinated ligands in the LSR structure. The additional electron-withdrawing effect should enhance the Lewis acidity of the metal ion, making the LSR useful for more weakly basic substrates. Substitution of fluorines for hydrogen also would decrease the number of interferring signals in the ^1H NMR spectrum.

The observation that the shifting power of variously fluorinated ligands (see **6**–**8** below) follows the order:[46]

$$Eu(fod)_3 > Eu(pfd)_3 > Eu(fhd)_3 > Eu(dpm)_3$$

is consistent with the inductive effect of fluorine. This effect accumulates and is felt even when the halogen is in the β and γ positions of the ligands. It was also found that the solubility order increased in the same direction. Burgett and Warner[46] also studied the effect to highly fluorinated ligands and found that larger shifts were observed for the more remote protons in substrates when $Eu(dfd)_3$ (**8** dfd = 1,1,1,2,2,6,6,7,7,7-decafluoro-3,5-dionato), rather than $Eu(fod)_3$ was used.

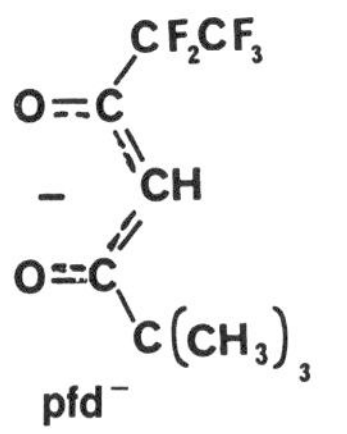

CF_2CF_3 / O=C / CH / O=C / $C(CH_3)_3$ — pfd⁻ **6**

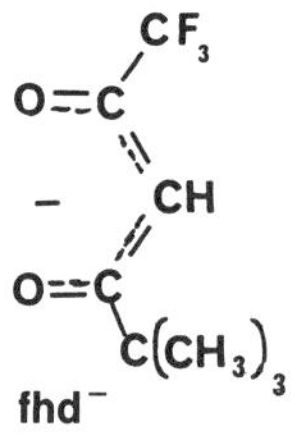

CF_3 / O=C / CH / O=C / $C(CH_3)_3$ — fhd⁻ **7**

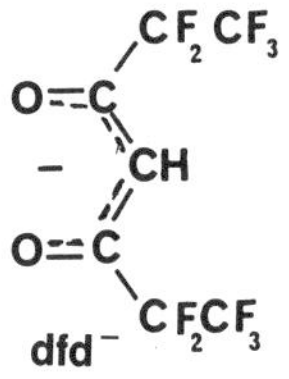

CF_2CF_3 / O=C / CH / O=C / CF_2CF_3 — dfd⁻ **8**

We have carried out studies in our lab with the even more highly fluorinated LSR, $Eu(tfn)_3$ (where tfn^- is 1,1,1,2,2,3,3,7,7,8,8,9,9,9-tetra decafluoro-4,6-nonandionato). The original purpose of our research was to induce significant LIS shifts for thioethers. Because thioethers are soft bases (compared to oxygen ethers) and the LSR reagents contain lanthanide ions that are hard acids, the thioether–LSR coordination complex forms only weakly.[24] We obtained shifts for thioethers using $Eu(tfn)_3$ and deduced the following LSR strength order:[47] $Eu(tfn)_3 >$ $Eu(fod)_3 > Eu(dpm)_3$ where

$$tfn^- = CF_3CF_2CF_2\overset{\overset{O}{\|}}{C}—\bar{C}H—\overset{\overset{O}{\|}}{C}CF_2CF_2CF_3$$

We also obtained reasonable LIS values with alkenes, nitro compounds,[48] and chloroalkyl compounds. A limitation of these reagents is their physical nature. The tfn LSR compounds proved difficult to dry and commercial samples often contained more than traces of the ligand precursor (Htfn = $CF_3CF_2CF_2COCH_2COCF_2CF_2CF_3$).

The most extensive work on highly fluorinated LSR compounds has been by R. E. Sievers et al.[15] Sievers has prepared and examined $Eu(dfhd)_3$, $Yb(dfhd)_3$, and $Pr(dfhd)_3$ where dfhd = 1,1,1,5,5,6,6,7,7,7-decafluoro-2,4-heptanedionato. It should be noted that the dfhd ligand, although isomeric with dfd, has one more α-fluorine than tfn. Perhaps, therefore, it is not surprising to find that LIS values from $Eu(dfhd)_3$ studies are of the order of magnitude of those caused by $Eu(tfn)_3$.

We now can suggest a grand order of shifting ability:

$$Eu(dfhd)_3 \cong Eu(tfn)_3 > Eu(fod)_3 > Eu(pfd)_3 > Eu(hfd)_3 > Eu(dpm)_3$$

Sievers has found that his $Ln(dfhd)_3$ reagents induce shifts in nitro-compounds and in nitriles. The nitro results were similar to ours using $Eu(tfn)_3$. The magnitudes of shifts for the nitro and nitrile compounds suggest the strong inherent Lewis acid strength for these LSR compounds.

The handling of Sievers' compound did involve some difficulty. Sievers reports that $Eu(dfhd)_3$ is isolated initially as the dihydrate $Eu(dfhd)_3(H_2O)_2$. He also suggests that despite published statements about handling, drying, and purification of $Eu(fod)_3$, this also is hydrated and is actually $Eu(fod)_3(H_2O)_{0.5}$.

Fluorination of the ligand effectively increases the solubility of the LSR.[49,51] A solution of 20 mg of 1-heptanol in 0.5 mL CCl_4 was able to dissolve 260 mg of $Eu(dfhd)_3$.

As mentioned earlier, LSR·S formation (eq. 1-1) normally involves rapid equilibration with free LSR and S. An exception is the complex between $Eu(fod)_3$ and dimethyl sulfoxide (DMSO). At a ratio of $[Eu(fod)_3]$: [DMSO] = 0.336, Evans and Wyatt[22] found at $-80°$ that complexed DMSO showed an ^1H NMR signal 3.42 ppm downfield of free DMSO. Reuben found that integration of the fod *t*-butyl group in this case was consistent with $Eu(fod)_3(DMSO)_2$ stoichiometry for the complex.[50]

x-Ray results for $Eu(dpm)_3$ complexes have been mentioned above. The hydrated complex of $Pr(fod)_3$ has been studied with x-rays and is apparently a dimer with $Pr(fod)_3$ units bridged by one of the waters of hydration:[52]

$$(fod)_3Pr\cdots \underset{H\ \ H}{O} \cdots Pr(fod)_3 \cdots H_2O$$

This complex, when evacuated at room temperature, loses one of the two waters of hydration; the bridged water is retained.[53,54]

Vapor-phase osmometry studies of $Ln(fod)_3$-type compounds show evidence for association. The number of $Ln(fod)_3$ units in the association aggregate depends on the solvent. In benzene, carbon tetrachloride, and cyclohexane, a high degree of self-association is observed. In the more polar solvent $CHCl_3$, monomeric $Ln(fod)_3$ units are observed; $Ln(dpm)_3$ appears to be monomeric in all solvents.[55]

References

1. Hinckley, C. C. *J. Am. Chem. Soc.*, **1969**, *91*, 5160.
2. Sanders, J. K. M.; Williams, D. H. *J. Am. Chem. Soc.*, **1971**, *93*, 641.
3. Mayo, B. C. *Chem. Soc. Rev.*, **1973**, *2*, 49.
4. Cockerill, A. F.; Davies, G. L. O.; Harden, R. C.; Rackham, D. M. *Chem. Rev.*, **1973**, *73*, 553.
5. Reuben, J. *Prog. NMR Spectrosc.*, **1973**, *9*, 1.
6. Hofer, O. In "Topics in Stereochemistry"; Allinger, N. L.; Eliel, E. L., Eds.; Wiley.: New York, 1976; Vol. 9. p. 111.
7. von Ammon, R.; Fischer, R. D. *Angew. Chem. Int. Ed. (Engl.)*, **1972**, *11*, 675.
8. Sievers, R. E. Ed. "Nuclear Magnetic Resonance Shift Reagents"; Academic Press: New York, 1973.
9. Sievers, R. E.; Brooks, J. J.; Cunningham, J. A.; Rhine, W. E. "Advances in Chemistry Series"; American Chemical Society: Washington, D.C., 1976; No. 150, p. 222.
10. Sullivan, G. R. In "Topics in Stereochemistry," Eliel, E. L.; Allinger, N. L., Eds.; Wiley: New York, 1976; Vol. 10, p. 287.
11. Williams, D. H. *Pure Appl. Chem.* **1974**, *40*, 25.
12. Petersen, M. R., Jr.; Wahl, G. H., Jr. *J. Chem. Ed.*, **1972**, *49*, 790.
13. Kuo, S. C.; Harriss, D. K.; Caple, R. *J. Chem. Ed.*, **1974**, *51*, 280.
14. Campbell, J. R. *Aldrichimica Acta*, **1971**, *4*, 55.
15. Kime, K. A.; Sievers, R. E. *Aldrichimica Acta*, **1977**, *10*, 54.
16. Wasson, J. R.; Lorentz, D. R. *Anal. Chem.*, **1976**, *48*, 250R.
17. Glasel, J. A. In "Current Research Topics in Bioinorganic Chemistry"; Lippard, S. J., Ed.; Wiley—Interscience: New York, 1973; p. 383.
18. Orrell, K. G. In "Annual Reports on NMR Spectroscopy"; Webb, G. A., Ed.; New York: Academic Press, 1979; Vol. 9, p. 1.
19. Inagaki, F.; Miyazawa, T. *Prog. NMR Spectrosc.*, **1981**, *14*, 67.
20. DeBoer, E.; Van Willigen, H. *Prog. NMR Spectrosc.* **1967**, *2*, 111.
21. Sanders, J. K. M.; Williams, D. H. *Chem. Commun.*, **1970**, 422.
22. Evans, D. F.; Wyatt, M. *Chem. Commun.*, **1972**, 312.
23. Pearson, R. *J. Am. Chem. Soc.*, **1963**, *85*, 3533.
24. Morrill, T. C.; Opitz, R. J.; Mozzer, R. *Tetrahedron Lett.*, **1973**, 3715.
25. DeMarco, P. V.; Elzey, T. K.; Lewis, R. B.; Wenkert, E. *J. Am. Chem. Soc.*, **1970**, *92*, 5743.
26. Rondeau, R. E.; Sievers, R. E. *J. Am. Chem. Soc.*, **1971**, *93*, 1522.
27. Eisentraut, K. J.; Sievers, R. E. *J. Am. Chem. Soc.*, **1965**, *87*, 5254.
28. Springer, C. S., Jr.; Meek, D. W.; Sievers, R. E. *Inorg. Chem.*, **1967**, *6*, 1105.
29. Sanders, J. K. M.; Hanson, S. W.; Williams, D. H. *J. Am. Chem. Soc.*, **1972**, *94*, 5325.

30. Crump, D. R.; Sanders, J. K. M.; Williams, D. H. *Tetrahedron Lett.*, **1970**, 4419.
31. Hawkes, G. E.; Liebfritz, D.; Roberts, D. W.; Roberts, J. D. *J. Am. Chem. Soc.*, **1973**, *95*, 1659.
32. Hawkes, G. E.; Marzin, C.; Liebfritz, D.; Johns, S. R.; Herwig, K.; Copper, R. A.; Roberts, D. W.; Roberts, J. D. In "Nuclear Magnetic Resonance Shift Reagents"; Sievers, R. E., Ed.; Academic Press: New York, 1973; p. 129.
33. Barry, C. D.; Dobson, C. M.; Sweigert, D. A.; Ford, L. E.; Williams, R. J. P. "Nuclear Magnetic Resonance Shift Reagents"; Sievers, R. E., Ed.; Academic Press: New York, 1973; p. 173.
34. Cramer, R. E.; Seff, K. *Chem. Commun.*, **1972**, 400; (b) *Acta Crystallogr.*, **1972**, *B28*, 3281.
35. Horrocks, W. deW. Jr. *J. Am. Chem. Soc.*, **1974**, *96*, 3022.
36. Shapiro, B. L.; Hlubucek, J. R.; Sullivan, G. R.; Johnson, L. F. *J. Am. Chem. Soc.*, **1971**, *93*, 3281.
37. Wolkowski, Z. W.; Beaute, C.; Jantzen, R. *Chem. Commun.*, **1972**, 619.
38. Sanders, J. K. M.; Williams, D. H. *Tetrahedron Lett.*, **1971**, 2813.
39. Witanowski, M.; Stefaniak, L.; Januszewski, H.; Wolkowski, Z. W. *Tetrahedron Lett.*, **1971**, 1653.
40. Gansow, O. A.; Loeffler, P. A.; Davis, R. E.; Willcott, M. R. III; Lenkinski, R. E. *J. Am. Chem. Soc.*, **1973**, *95*, 3390.
41. Tori, K.; Yoshimura, Y. *Tetrahedron Lett.*, **1973**, 3127.
42. Barry, C. D.; North, C. T.; Glasel, J. A.; Williams, R. J. P.; Xavier, A. V. *Nature*, (London) **1971**, *232*, 236.
43. Levine, B. A.; Williams, R. J. P. *Proc. Roy. Soc. London*, **1975**, *A245*, 5.
44. Smith, G. V.; Boyd, W. A.; Hinckley, C. C. *J. Am. Chem. Soc.*, **1971**, *93*, 6319.
45. DeHorrocks, W., Jr.; Sipe, K. P.; Luber, J. R. *J. Am. Chem. Soc.*, **1971**, *93*, 5258.
46. Francis, H. E.; Wagner, W. F. *Org. Mag. Reson.*, **1972**, *4*, 189. Burgett, C. A.; Warner, P. *J. Mag. Reson.*, **1972**, *8*, 87.
47. Morrill, T. C.; Clark, R. A.; Bilobran, D.; Youngs, D. S. *Tetrahedron Lett.*, **1975**, 397.
48. Morrill, T. C.; Furman, B. Unpublished work.
49. Dyer, D. S.; Cunningham, J. A.; Brooks, J. J., Sievers, R. E.; Rondeau, R. E., In "Nuclear Magnetic Resonance Shift Reagents"; Sievers, R. E., Ed.; Academic Press: New York, 1973; p. 27.
50. Reuben, J., *Prog. NMR Spectroscopy*, **1973**, *9*, 1.
51. Richardson, M. F.; Sievers, R. E. *Inorg. Chem.*, **1971**, *10*, 498.
52. DeVilliers, J. P. R.; Boeyens, J. C. A. *Acta Crystallogr.*, **1971**, *B27*, 692.
53. Dyer, D. S.; Cunningham, J. A.; Brooks, J. J.; Sievers, R. E.; Rondeau, R. E. In "Nuclear Magnetic Resonance Shift Reagents"; Sievers, R. E., Ed.; Academic Press: New York, 1973; p. 21.
54. Springer, C. S.; Bruder, A. H.; Tanny, S. R.; Pickering, M.; Rockefeller, H. A. In "Nuclear Magnetic Resonance Shift Reagents"; Sievers, R. E., Ed.; Academic Press: New York, 1973; p. 283.
55. Inagaki, F.; Miyazawa, T.; Miwa, M.; Saito, H.; Sugimura, T. S. *Biochem. Biophys. Res. Commun.*, **1978**, *85*, 415.

2

COMPUTER APPROACHES TO THE GEOMETRY OF THE LSR–SUBSTRATE COMPLEX: GEOMETRY OF SHIFT REAGENT–SUBSTRATE COMPLEXES, SIMULATION OF INDUCED CHEMICAL SHIFTS

Julien B. Chiasson and Krzysztof Jankowski*

Département de Chimie,
Université de Moncton,
Moncton, N.B. E1A 3E9 Canada

Introduction

The use of NMR shift reagents is now standard in the study of molecular structures. The NMR spectra, simplified by the use of these shift reagents, are becoming more legible. Therefore, easier interpretation of the spectra and easier conformational and configurational studies have become possible. The shift reagents, in most cases, are lanthanide *tris*-β-diketonates, act as Lewis acids, and produce complexation with

* To whom all inquiries should be addressed.

Morrill (ed): LANTHANIDE SHIFT REAGENTS IN STEREOCHEMICAL ANALYSIS

the substrates studied. The latter bear a Lewis base referred to as the donor group (eg, OH, NR, C=O, etc).

The induced shifts arise from dipolar (pseudocontact) interactions between the shift reagent and the substrate. Such a shift is a measure of the time-average environment of the nuclei in the complexed and noncomplexed forms. The value of the induced chemical shift therefore depends on the stability of the complex, the quantity of shift reagent used, and the value of the induced shifts for the given complex. Because the induced shifts are the most important in the elucidation of the molecular geometry, this chapter concentrates on their prediction and simulation by a computer. This technique is illustrated by some examples.

There is a danger involved in working with computers, especially via a conversational mode. We start by asking *ourselves* questions, and we end up providing answers to these questions. In spite of this evident "introduction to alienation" we try to preserve the character of this operation in this chapter. A dialog with ourselves is, after all, the essence of research.

Origin of the Induced Shift

When the substrate is mixed with shift reagents, the shifts in the NMR signals are caused by two different phenomena: contact interaction and dipolar interaction.

Contact Interaction

Contact interactions (also called Fermi interactions) occur when the electron density (as determined by unpaired electrons) of the shift reagent is nonzero for the nucleus under observation. The chemical shift of the nucleus undergoes a modification: the shift depends on the electron density around the nucleus when the local magnetic field is perturbed.

Dipolar Interaction

If an unpaired electron is highly localized on the metal ion, the contact interaction becomes weak or zero. The magnetic anisotropy generated in the neighboring space by the local dipolar magnetic field is then the only significant contribution to the induced shift. The nuclei situated in this field experience different values of magnetic field. The chemical shifts of the nuclei become displaced in a nonuniform manner. These displacements are the dipolar induced shifts, often called just dipolar shifts.

The dipolar shifts for lanthanides are described by the McConnell and Robertson equation:[1]

$$\Delta_i = K(3 \cos^2 \theta_i - 1)/r_i^3 \tag{2-1}$$

where Δ_i is the dipolar shift, K is a group of constants, r_i is the distance between the ith nucleus and the paramagnetic ion, and θ_i is the angle between the vector r_i and the magnetic axis of the paramagnetic ion.* The equation is a particular case of a set of more complex ones.[1] The simplification owes to the effective axiality of the lanthanide ion magnetic field, explained by different mechanisms.[3,15–35] The assumption of axiality is supported by experiments. It has been found that the magnetic axis is virtually colinear with the paramagnetic ion–donor group vector[3,15,22,36–46].

Utilization of the Predicted Induced Shifts

Simulation of induced shifts gives a rationale for interpreting NMR spectra. The attribution of the peaks then becomes much simpler. Consider two protons such that their unshifted signals (ie, chemical shifts) are expected to be very close to each other. If their induced shifts are predicted to be different, then it becomes an easy task to identify which signal is which by comparing the relative sizes of their actual (observed) induced shifts.

The simulation also gives a way to predict whether the different proposed structures for a given compound possess characteristics sufficiently distinctive so that the shift reagent technique can be applied. A classic example is the one given by Hawkes, Roberts et al,[3] namely, borneol and isoborneol. It is easy to see even before an experiment is started that the two isomers are differentiable by the shift reagent technique. In isoborneol, where the hydroxy group is *exo*, the OH projects outwardly from the rigid norbornane skeleton; in borneol, where the hydroxy is *endo*, it is tucked under the skeleton. A second application

* Atomic coordinates are discussed in refs. 2–14.

is the identification of a compound structure by comparing the observed values of the induced shifts with those computed for a set of proposed structures. This can be a continuation of the previous application, if the conclusion reached is that a differentiation is possible.

The combined technique just described, simulation–titration–identification, is applicable in studies of different kinds of isomerisms: positional, geometric (cis–trans, syn–anti, or axial–equatorial, the latter mostly for rigid systems), and to a lesser extent rotational and fluxional conformers.[47]

On the pedagogic side, the relation between the spectroscopic properties and the molecular geometries (here the Cartesian coordinates of the atoms) permits a better understanding of the third dimension and of complete molecular structures. Moreover this has made the induced shift a much less abstract quantity.

Cartesian Coordinates

In order to perform the simulation of the induced shifts, a way of expressing the geometry of the complex is needed. This is best done by using the Cartesian coordinate reference frame. The coordinates can be obtained in different ways: x-ray measurements, geometric computations,[10,14,48–52] measurements on molecular models, or a combination of these three methods.[6,53] In the third case the data can be obtained by a simple method[2]—by using a coordinate-measuring instrument. Two different types are available: sonic (eg, SAC GP-2-3D) or displacement reading (eg, Mitutoyo coordinate-measuring instrument). We made our measurements on Dreiding models (scale 5 cm/Å or 2.5 cm/Å) using the Mitutoyo instrument. The results obtained in our computations confirm that, although their use has been severely criticized, the Dreiding models are still good approximations of the real state of the molecules in solution. Moreover, the Corey[54] vector analysis regains its rationale, especially in its pedagogical aspect. The average bond lengths and angles lead to a computation of the molecular geometry parameters applicable in particular to NMR spectroscopy; eg, dihedral angles and interatomic distances lead to coupling constants in Karplus-type calculations. Usually the measurement errors in the coordinates are negligible[18] compared to the approximations arising from the utilization, for instance, of mean hybridization angles.

A more accurate study of the effect of atomic coordinate errors on the lanthanide position optimization has been made by Hinckley and Brumley.[55]

Position of the Lanthanide Ion in the Complex or How to Place the Lanthanide with Respect to Different Donor Groups

Now that we have discussed how to measure the molecular geometry data for the substrate molecule, we still need to know how this relates to the geometry of the complex, that is, how the molecule is positioned with respect to the paramagnetic center of the shift reagent complex. Another way to look at the problem is to ask where the lanthanide ion is situated with respect to the molecule. This is best answered by locating the central ion in a place having a chemical significance, ie, near a Lewis base functional group, more precisely in the axis of a lone-pair orbital of a donor atom. When there is more than one lone-pair orbital, or more than one donor in the group, the induced shifts for each of the possible lanthanide positions can be computed and a weighted mean value found for each proton. Likewise, a most probable unique position can be chosen for the lanthanide and the computations done for that position only. In both these cases, the steric hindrance and the molecular symmetry must be taken into account (a bit of chemical common sense is welcome here, too).

The next section provides a review of some of the results concerning the location of the lanthanide ion with respect to certain donor groups. For reasons of simplicity, and also to follow the general trend in the literature, the position of the lanthanide is given in spherical coordinates (R, ϕ, Ω) as explained in the organigram of the program SIMULATION provided in the section, "Program SIMULATION: Structure and Usage."

Carboxylic Group[56]

Both oxygens forming the carboxylic group are considered to be complexed equally with the lanthanide ion. Therefore, a good approximation is obtained in our models by placing the paramagnetic center on the axis bisecting the angle O–C–O and in the plane of the carboxylic group. For our simulation program, the position of the donor can be given as the midpoint between the two oxygens. Taking the carboxylic carbon as the program's first reference, the position of the lanthanide is (2.2 Å, 180°, 0°) (Figure 2-1).

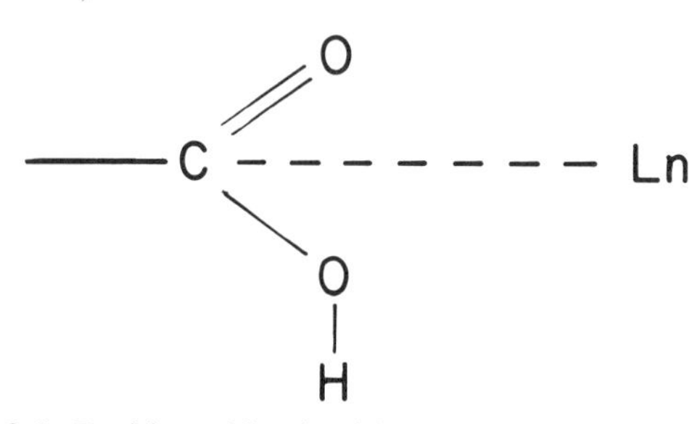

Figure 2-1. Position of lanthanide in carboxylic acid complex.

Amines,[17] *Alcohols,*[17,18,57–61] *and Some Alkoxyl compounds*[60]

Following the principles of free orbitals, the values of R and Ω are roughly 2.7 Å and 135°, respectively. The value of ϕ depends on the steric hindrance and the molecular geometry. In the case of tertiary alkyl groups, three contributing rotamers, with possibly different populations, can be expected. If the tertiary carbon is considered the first reference and an adjacent atom to be the second reference, then the three rotamers have the angle Ω of 60°, 180°, and 300°, respectively, for the position of the lanthanide. Depending on the symmetry directly in the neighborhood of the C—X bond, some ordering in the value for the different populations can be expected. For example, if there is an axis of symmetry passing through the C—X bond then all three populations can be assumed to be roughly the same. If there is only a plane of symmetry, then two of the rotamers will have the same population; the other one, located in the plane of symmetry, will have a higher or lower population depending on the steric hindrance. If there is no symmetry around the C—X bond then the populations are estimated by the degree of steric hindrance for each rotamer. The analyses follow essentially the same pattern for both secondary and primary cases.

Pyridine and Other Tertiary Heterocyclic Bases[62]

If in tertiary heterocyclic bases the middle point between the two carbons attached to the nitrogen atom is taken as the first reference point, and one of these carbons as the second reference point, then if the molecule is symmetrical, the position of the lanthanide is (2.5 Å, 180°, 0°). Otherwise, a more probable position is either (2.5 Å, 138°, 0°) or (2.5 Å, 138°, 180°), depending on which side is less hindered (Figure 2-2).

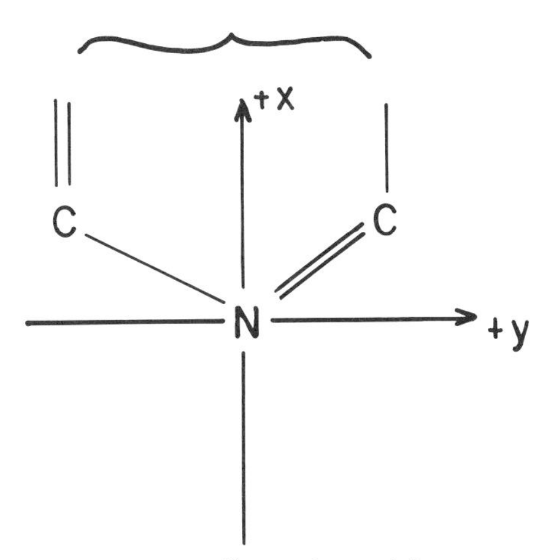

Figure 2-2. Coordinates for pyridine system.

Aldehydes,[63] Ketones,[60,63–67] Lactams,[36] and Lactones[68]

In the aldehydes, ketones, lactams, and lactones the lanthanide can be located in two different positions, (2.5 Å, 140°, 0°) and (2.5 Å, 140°, 180°), if the two indicated (A_1, A_2) atoms (see Figure 2-3) are used as reference points. The populations of the two complexes are determined by the steric hindrance in proximity to the donor group. (Figures 2-3 and 2-4).

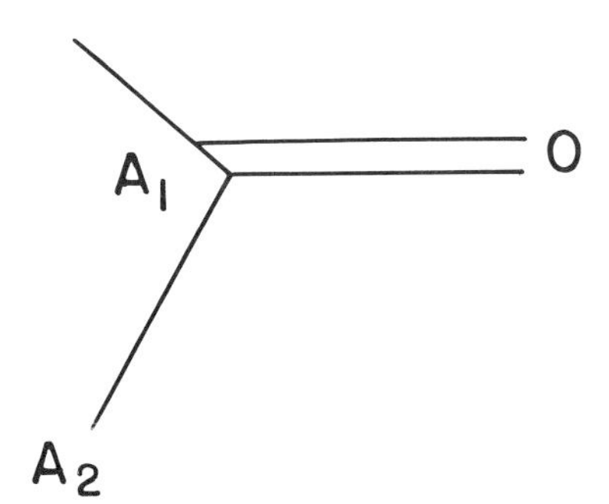

Figure 2-3. Reference atoms (A_1, A_2) for aldehyde, ketone, lactam and lactone complexes.

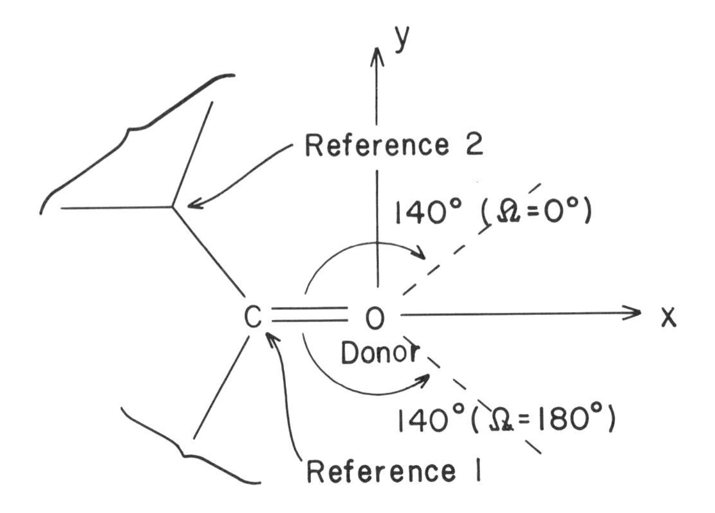

Figure 2-4. Coordinates for complexes of carbonyl compounds (see Figure 2-3). The z axis is perpendicular to the xy plane.

Cyclic Ether and Epoxides[69]

If in cyclic ether and epoxides the midpoint between the two carbons directly attached to the oxygen and one of these carbons, respectively, are taken as the first and second reference point, then the lanthanide may assume the following locations: (2.5 Å, 138°, 90°) and (2.5 Å, 138°, 270°). Even though there is no study on ethers that evaluates the contributions with respect to the free orbital principle, Caple and co-workers'[69] results indicate that the less hindered side is favored for other groups.

Program SIMULATION: Structure and Usage

We have written a set of subroutines permitting the computation of the relative induced shift for a compound. The inputs needed are the geometry of the molecule and the geometry of the molecule–shift reagent complex.

What Kind of Induced Shifts Are We Computing?

It has been shown that for lanthanide shift reagents, in ^{1}H NMR, dipolar interactions are the predominant—if not the exclusive—contribution to the induced shifts. Therefore, our program computes the shift as if it has resulted only from the dipolar term. In the ^{13}C NMR shift reagent experiment, the dipolar contribution is predominant if the carbon is removed by at least three bond lengths from the donor.

Because we are interested mainly in ^{1}H NMR spectra, the geometry of the molecule is described by a set of Cartesian coordinates for the protons. To describe the geometry of the complex it is necessary to designate a donor, ie, a site of complexation. Also, the lanthanide must be located with respect to the donor. To make the task easier, we define an internal Cartesian reference frame in which the donor is, by definition, the origin. Two other points serve to define, respectively, the x axis and the xy plane (Figures 2-5–2-7).

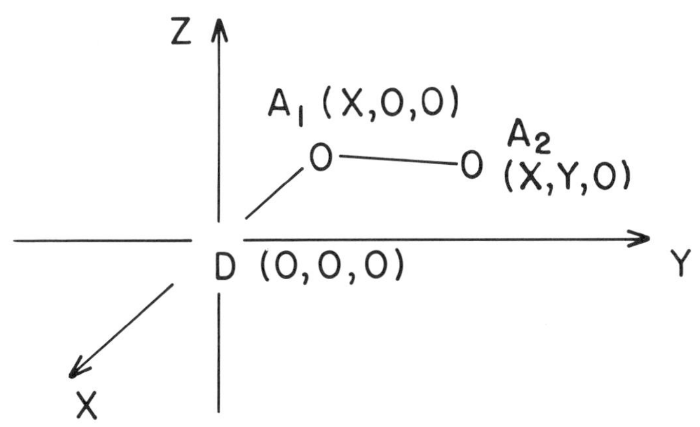

Figure 2-5. Cartesian coordinates for LSR complex. D = donor, A_1 defines x-axis and A_2 the xy plane.

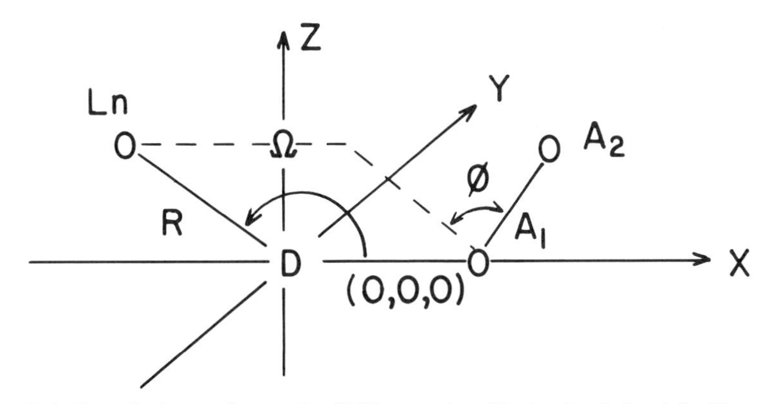

Figure 2-6. Spherical coordinates for LSR complex (D, A_1, A_2 defined in Figure 2-5).

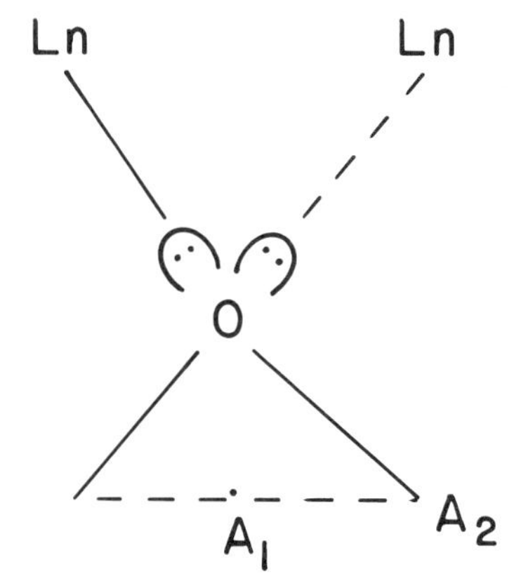

Figure 2-7. Donor (O) with two donor sites.

How To Define a New Frame of Reference and Why

Our program is so written that the donor is considered the origin of a Cartesian system, the first reference point is on the negative side of the x axis, and the second reference point on the +y region of the xy plane. This program design approach has a twofold effect. First, simpler subroutines can be written by diminishing the number of parameters passing from one part of the program to the other: the donor is always thought to be at the origin. Even if two donors are considered, each one still can be thought of as at the "origin" of its respective "molecule." This can be done by entering the position of the protons twice but giving different reference points for each set.

The second effect constitutes a simplification for the user. If he or she wants to enter the Cartesian coordinates of the lanthanide, it is easier to see the position with respect to the origin than with respect to an arbitrary point.

The main consideration in choosing the three points serving as reference points is that they should *not* be colinear. In the section above entitled "Position of the Lanthanide Ion in the Complex . . . ," suggestions are given for the reference points for specific compounds. Now a spherical coordinate system can be used, superimposed on the internal Cartesian reference frame, to describe the position of the lanthanide with respect to the molecule. The "triplet" (R, ϕ, Ω) is called the spherical coordinates of the lanthanide. Here R is the distance between the lanthanide and the donor; ϕ is an elevation angle with respect to the minus side of the x axis, ie, a bond angle; and Ω is the rotation angle with respect to the xy side of the xy plane (a torsional or dihedral angle)

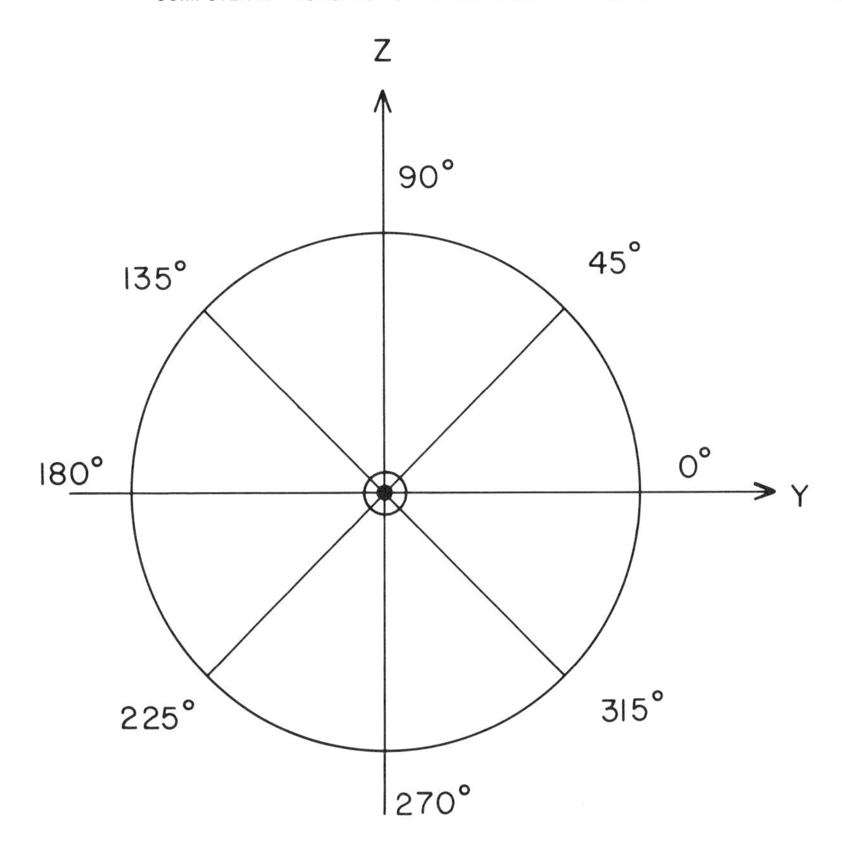

Figure 2-8. Cartesian coordinates used to visualize the rotation angle, Ω, with respect to the xy side of the xy plane. The +x axis is pointing outward.

(Figure 2-8). For example, the position

$$(x, y, z) = (2.5, 0, 0)$$

is equivalent to

$$(R, \phi, \Omega) = (2.5, 180, \text{anything})$$

and

$$(x, y, z) = \left(\frac{3\sqrt{2}}{2}, \frac{3\sqrt{2}}{2}, 0\right)$$

is equivalent to

$$(R, \phi, \Omega) \quad (3, 135, 0)$$

Comments on Programming and an Example

In order to illustrate the practical use of our programs, we have chosen the example of a dihydropyran derivative, ethyl 2-methyl-5,6-dihydro-α-pyran 6,6-dicarboxylate. This unsaturated, sugar-related compound can exist in solution in one of two conformations, represented in (Figure 2-9).

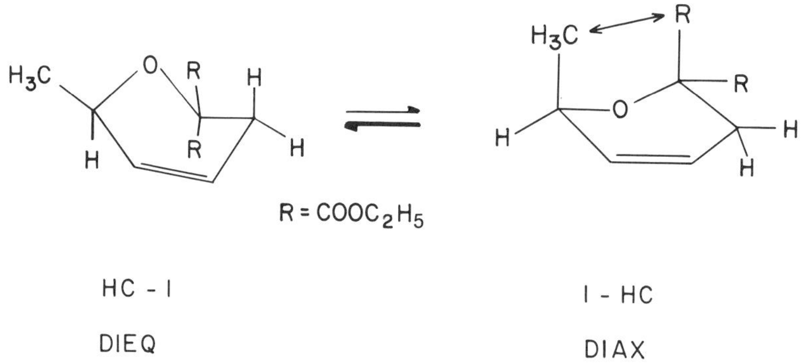

$R = COOC_2H_5$

HC - I

DIEQ

I - HC

DIAX

Figure 2-9. Conformations of ethyl 2-methyl-5,6-dihydro-α-pyran 6,6-dicarboxylate.

Although the 1,3-diaxial interaction shown above seems to affect the HC-1 $\rightleftharpoons$ 1-HC equilibrium, we have assumed the possibility of the presence of both conformations [called here DIAX (1-HC) and DIEQ (HC-1)]. The conformational analysis of similar molecules has been accomplished previously by our group.[70]

Let us now analyze in terms of this example our three main programs: SIMULATION, PGM1, and PGM2.

The program SIMULATION involves the simplest possible use of the subroutines. It takes as input the molecular geometry of the substrate–lanthanide complex and returns a relative induced shift for each of the protons given (see examples 1 and 2).

The program PGM1 is somewhat more complex. It takes as input two possible geometries for the complex, and it returns relative weighted induced shifts for different weightings. For instance, in examples 3, 4, and 5 two conformations of the molecule are used, one in which the methyl is axial (DIAX) and the other in which it is equatorial (DIEQ). The molar fraction is that of the first form entered that contributes to the induced shifts.

The last program, PGM2 (example 6), finds the ranges for the induced shifts if the lanthanide is allowed to move in a cube centered on a given position.

ADJUST

There exists one subroutine such that the user may (and probably will) want to modify; it is called ADJUST. This subroutine permits the user to average the signal of two (or more) protons (eg, protons of a methyl group) and to scale the resulting shifts so that the largest one is ~10 ppm. The first line of the subroutine, as given, performs the averaging of the signals of the last six "protons." In fact those six "proton" positions are the positions of the three protons of the methyl in our "example." Now the question arises: why six positions for three protons? There are two ways of computing the induced shift of a mobile proton (eg, methyl protons). The first is to take a number of positions allowable for the proton,[8] compute the induced shifts for each position, and then find the mean induced shift. We chose six points on the circle of rotation of the protons of the methyls, three for the protons in the eclipsed form and three in the staggered form, and these three protons are considered equivalent. The second way is to find the average position,[9] and then to compute the induced shift for that position only.

If the methyl, for example, is near the complexation site, it is preferable to use the first scheme; note that the variations from one point to another on the rotation circle may be important. If, in contrast, the methyl is removed from the complexion site, then the maximum difference in the value for a proton is not large; the induced shift can then be approximated by using a mean point. Even in such a case, it may still be preferable to use the first scheme to avoid less defined zones.

Of course, the user can write his or her own program calling for our subroutines. For example, he or she can implement the algorithm described by ApSimon and Beierbeck.[71]

Example 1: Simulation of the Induced Shifts for DIAX (1-HC). Methyl group protons are represented by six rotamer positions, 6, 7, 8, 9, 10 and 11:

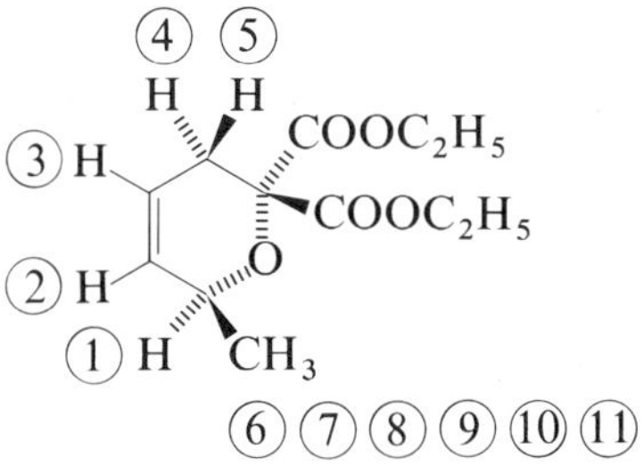

(4) (5)
H H
(3) H
$COOC_2H_5$
$COOC_2H_5$
O
(2) H
(1) H CH_3
(6) (7) (8) (9) (10) (11)

```
        DIAX
  0.175      1.015     0.397
 −1.243      1.628    −1.485
 −3.162      0.275    −1.863
 −3.137     −1.499     0.079
 −2.853     −2.09     −1.568
  1.212     −1.172    −1.233
  0.814     −0.312    −2.206
  1.879      0.44     −0.751
  0.763     −0.933    −1.877
  1.867     −0.391    −0.49
  1.309      0.762    −1.78

        O1DIAX
 −0.534     −0.901     0.414

        C6DIAX
 −1.07      −1.845    −0.47

        C7DIAX
 −1.112     −3.3       0.283

        O8DIAX
 −1.779     −4.212    −0.105

        O9DIAX
 −0.398     −3.369     1.508

        C12DIAX
 −0.198     −2.119    −1.667

        O13DIAX
  0.964     −1.889    −1.629

        O14DIAX
 −0.954     −2.519    −2.712

        SIMULATION
HOW MANY PROTONS IN YOUR SYSTEM?
□:
        11
GIVE THEIR POSITIONS (A N×3 MATRIX)
□:
        DIAX
GIVE THE POSITION OF THE DONOR
□:
        (O8DIAX+O9DIAX+O13DIAX+O14DIAX)÷4
GIVE THE POSITION OF THE FIRST REFERENCE POINT
□:
        C6DIAX
GIVE THE POSITION OF THE SECOND REFERENCE POINT
□:
        O1DIAX


THE REFERENCE POINTS IN THE NEW COORDINATE SYSTEM
DONOR                  :     .00000     .00000    .00000
FIRST  REFERENCE POINT:   −1.29487     .00000    .00000
SECOND REFERENCE POINT:   −2.09680    1.14751    .00000
```

POSITIONS OF THE PROTONS IN THE NEW COORDINATE SYSTEM

PROTON	X	Y	Z
1	−3.50906	2.05484	−1.16462
2	−4.24859	−.06986	−2.09584
3	−3.75027	−2.02220	−.83248
4	−2.55815	−1.14076	1.34030
5	−1.57995	−2.07947	.19815
6	−.80693	1.37276	−2.02979
7	−1.53582	.63539	−2.90714
8	−2.06773	2.50063	−2.67326
9	−1.07123	.67883	−2.37917
10	−1.38647	2.48857	−2.13052
11	−2.37661	1.49072	−3.28326

IS THE LANTHANIDE POSITION GIVEN IN CARTESIAN (C)
OR SPHERICAL (S) COORDINATES?
S
ENTER THE LANTHANIDE POSITION
RADIUS OMEGA PHI
□:
2.5 180 0
THE CARTESIAN COORDINATES OF THE LANTHANIDE ARE:
(X,Y,Z) = 2.500 .0000 .0000

PROTON	INDUCED SHIFT
1	4.15360
2	3.43982
3	4.02928
4	7.63056
5	10.00000
6	5.53857

Example 2: Simulation of the Induced Shifts for DIEQ (HC-1)

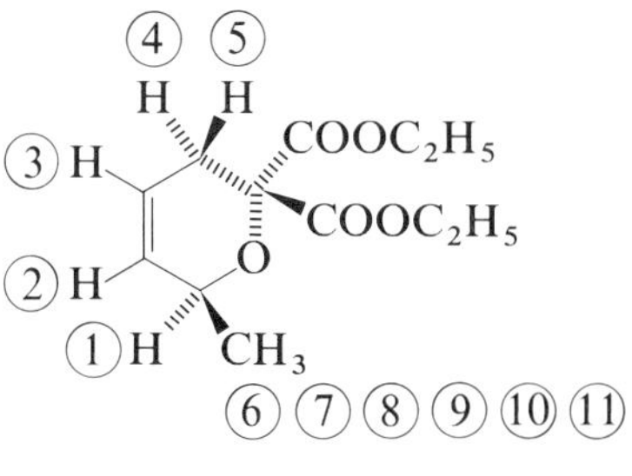

(4) (5)
H H
(3) H
$COOC_2H_5$
$COOC_2H_5$
O
(2) H
(1) H
CH_3
(6) (7) (8) (9) (10) (11)

DIEQ

−2.627	−1.304	0.545
−2.917	0.364	2.513
−1.446	2.189	1.969
0.007	1.904	0.059
0.817	0.98	1.322
−1.79	−3.135	1.945
−3.132	−2.22	2.791
−1.419	−1.952	3.272
−2.25	−1.769	3.383
−2.822	−2.912	2.176
−1.068	−2.657	2.51

```
        O1DIEQ
-0.559      -1.286      0.802

        C6DIEQ
-0.292      -0.224     -0.103

        C7DIEQ
 0.993      -0.556     -0.933

        O8DIEQ
 1.293      -1.744     -1.079

        O9DIEQ
 1.64        0.554     -1.335

        C12DIEQ
-1.453      -0.014     -1.125

        O13DIEQ
-1.733       1.138     -1.44

        O14DIEQ
-2.026      -1.191     -1.516

        SIMULATION
HOW MANY PROTONS IN YOUR SYSTEM?
□:
        11
GIVE THEIR POSITIONS (A N×3 MATRIX)
□:
        DIEQ
GIVE THE POSITION OF THE DONOR
□:
        (O8DIEQ+O9DIEQ+O13DIEQ+O14DIEQ)÷4
GIVE THE POSITION OF THE FIRST REFERENCE POINT
□:
        C6DIEQ
GIVE THE POSITION OF THE SECOND REFERENCE POINT
□:
        O1DIEQ

THE REFERENCE POINTS IN THE NEW COORDINATE SYSTEM
DONOR                   :      .00000     .00000    .00000
FIRST  REFERENCE POINT:     -1.24547     .00000    .00000
SECOND REFERENCE POINT:     -2.09049    1.14197    .00000

POSITIONS OF THE PROTONS IN THE NEW COORDINATE SYSTEM
PROTON         X             Y            Z
```

PROTON	X	Y	Z
1	−1.97543	1.52369	2.04550
2	−4.07009	.04995	2.46959
3	−3.55483	−2.04102	1.39591
4	−1.53439	−2.13429	.07317
5	−2.67137	−1.30480	−.98714
6	−3.18373	3.24616	.78603
7	−4.18153	2.64111	2.19748
8	−4.56130	2.09154	.52676
9	−4.74156	2.07023	1.36547
10	−3.50000	3.22910	1.81937
11	−3.72975	2.67654	.11790

IS THE LANTHANIDE POSITION GIVEN IN CARTESIAN (C)
OR SPHERICAL (S) COORDINATES?
S
ENTER THE LANTHANIDE POSITION
RADIUS OMEGA PHI
□:
 2.5 180 0
THE CARTESIAN COORDINATES OF THE LANTHANIDE ARE:
(X,Y,Z) = 2.5000 .0000 .0000

PROTON	INDUCED SHIFT
1	6.54976
2	3.33416
3	3.97632
4	10.00000
5	7.66269
6	2.81686

Example 3: Simulation of Induced Shifts for Different Equilibria between DIEQ (HC-1) and DIAX (1-HC), for PGM1

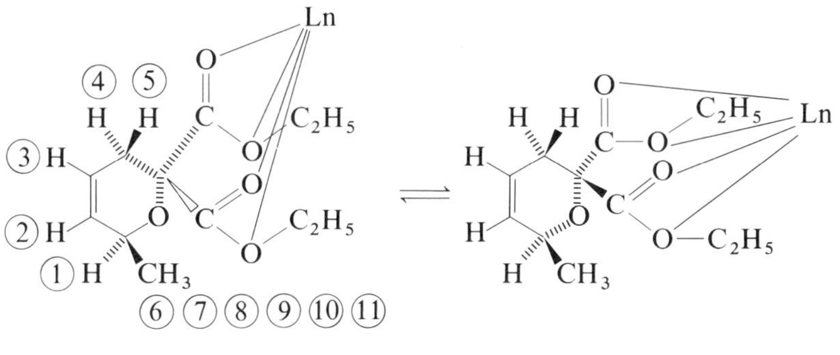

PGM1

DATA FOR THE FIRST FORM
HOW MANY PROTONS IN YOUR SYSTEM?
□:
 11
GIVE THEIR POSITIONS (A N×3 MATRIX)
□:
 DIEQ
GIVE THE POSITION OF THE DONOR
□:
 (O8DIEQ+O9DIEQ+O13DIEQ+O14DIEQ)÷4
GIVE THE POSITION OF THE FIRST REFERENCE POINT
□:
 C6DIEQ
GIVE THE POSITION OF THE SECOND REFERENCE POINT
□:
 O1DIEQ

```
THE REFERENCE POINTS IN THE NEW COORDINATE SYSTEM
DONOR                  :     .00000    .00000   .00000
FIRST  REFERENCE POINT:   -1.24547    .00000   .00000
SECOND REFERENCE POINT:   -2.09049   1.14197   .00000
```

POSITIONS OF THE PROTONS IN THE NEW COORDINATE SYSTEM

PROTON	X	Y	Z
1	−1.97543	1.52369	2.04550
2	−4.07009	.04995	2.46959
3	−3.55483	−2.04102	1.39591
4	−1.53439	−2.13429	.07317
5	−2.67137	−1.30480	−.98714
6	−3.18373	3.24616	.78603
7	−4.18153	2.64111	2.19748
8	−4.56130	2.09154	.52676
9	−4.74156	2.07023	1.36547
10	−3.50000	3.22910	1.81937
11	−3.72975	2.67654	.11790

```
IS THE LANTHANIDE POSITION GIVEN IN CARTESIAN (C)
OR SPHERICAL (S) COORDINATES?
S
ENTER THE LANTHANIDE POSITION
RADIUS OMEGA PHI
⎕:
      2.5  180  0
THE CARTESIAN COORDINATES OF THE LANTHANIDE ARE:
(X,Y,Z) = 2.5000    .0000    .0000
```

```
DATA FOR THE SECOND FORM
HOW MANY PROTONS IN YOUR SYSTEM?
⎕:
      11
GIVE THEIR POSITIONS (A N×3 MATRIX)
⎕:
      DIAX
GIVE THE POSITION OF THE DONOR
⎕:
      (O8DIAX+O9DIAX+O13DIAX+O14DIAX)÷4
GIVE THE POSITION OF THE FIRST REFERENCE POINT
⎕:
      C6DIAX
GIVE THE POSITION OF THE SECOND REFERENCE POINT
⎕:
      O1DIAX
```

```
THE REFERENCE POINTS IN THE NEW COORDINATE SYSTEM
DONOR                  :     .00000    .00000   .00000
FIRST  REFERENCE POINT:   -1.29487    .00000   .00000
SECOND REFERENCE POINT:   -2.09680   1.14751   .00000
```

POSITIONS OF THE PROTONS IN THE NEW COORDINATE SYSTEM

PROTON	X	Y	Z
1	−3.50906	2.05484	−1.16462
2	−4.24859	−.06986	−2.09584
3	−3.75027	−2.02220	−.83248
4	−2.55815	−1.14076	1.34030
5	−1.57995	−2.07947	.19815
6	−.80693	1.37276	−2.02979
7	−1.53582	.63539	−2.90714
8	−2.06773	2.50063	−2.67326
9	−1.07123	.67883	−2.37917
10	−1.38647	2.48857	−2.13052
11	−2.37661	1.49072	−3.28326

IS THE LANTHANIDE POSITION GIVEN IN CARTESIAN (C)
OR SPHERICAL (S) COORDINATES?
C
ENTER THE LANTHANIDE POSITION
X Y Z
□:
2.5 0 0

RELATIVE INDUCED SHIFTS

MOLAR FRACTION	1	2	3	4	5	ME
(FIRST FORM)						
.00	10.00	8.28	9.70	18.37	24.08	13.33
.10	10.00	7.81	9.16	17.91	22.23	11.99
.20	10.00	7.38	8.67	17.49	20.58	10.78
.30	10.00	6.99	8.24	17.12	19.08	9.69
.40	10.00	6.65	7.84	16.78	17.73	8.70
.50	10.00	6.33	7.48	16.47	16.50	7.81
.60	10.00	6.04	7.15	16.19	15.38	6.99
.70	10.00	5.77	6.85	15.93	14.34	6.23
.80	10.00	5.53	6.57	15.69	13.39	5.54
.90	10.00	5.30	6.31	15.47	12.51	4.90
1.00	10.00	5.09	6.07	15.27	11.70	4.30

Example 4: Simulation of Induced Shifts for Different Equilibria of DIAX (1-HC) Conformation with Lanthanide Positioned at Axial or Equatorial Carboethoxy Group, for PGM1

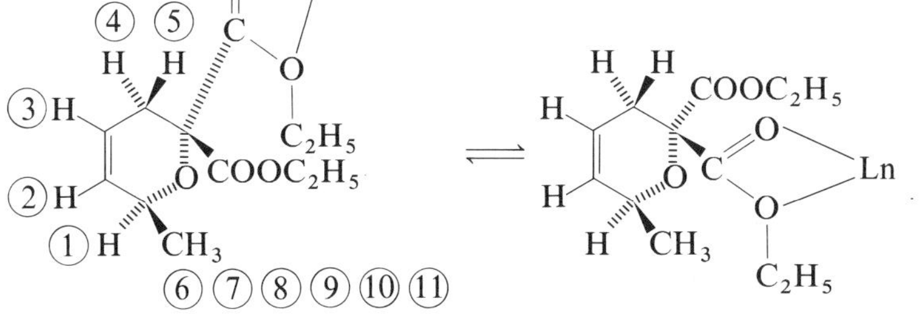

Ln
O
C
O
C_2H_5
(4) (5)
H H
(3) H
(2) H
(1) H
$COOC_2H_5$
CH_3
(6) (7) (8) (9) (10) (11)
⇌
H H
H
H
H
$COOC_2H_5$
O
C
Ln
O
CH_3
C_2H_5

```
        PGM1

DATA FOR THE FIRST FORM
HOW MANY PROTONS IN YOUR SYSTEM?
□:
        11
GIVE THEIR POSITIONS (A N×3 MATRIX)
□:
        DIAX
GIVE THE POSITION OF THE DONOR
□:
        (O8DIAX+O9DIAX)÷2
GIVE THE POSITION OF THE FIRST REFERENCE POINT
□:
        C7DIAX
GIVE THE POSITION OF THE SECOND REFERENCE POINT
□:
        O1DIAX

THE REFERENCE POINTS IN THE NEW COORDINATE SYSTEM
DONOR                   :        .00000     .00000   .00000
FIRST    REFERENCE POINT:       −.64520     .00000   .00000
SECOND REFERENCE POINT:       −2.36296    1.77644   .00000

POSTIONS OF THE PROTONS IN THE NEW COORDINATE SYSTEM
PROTON         X              Y             Z
------------------------------------------------
         1    −3.80476      3.19917     −.26662
         2    −5.54316      1.74586      .62461
         3    −4.82965     −.04362      2.01913
         4    −2.22045      .23503      2.20209
         5    −2.82911    −1.18069      1.32585
         6    −3.16165      1.08480    −2.17217
         7    −4.46106      .78845     −1.91057
         8    −4.05020      2.65510    −2.31501
         9    −3.77742      .61855     −1.89487
        10    −3.24960      2.32238    −2.39957
        11    −4.98320      1.92642    −2.02181

IS THE LANTHANIDE POSITION GIVEN IN CARTESIAN (C)
OR SPHERICAL (S) COORDINATES?
C
ENTER THE LANTHANIDE POSITION
X  Y  Z
□:
        2.5  0  0

DATA FOR THE SECOND FORM
HOW MANY PROTONS IN YOUR SYSTEM?
□:
        11
GIVE THEIR POSITIONS (A N×3 MATRIX)
□:
        DIAX
GIVE THE POSITION OF THE DONOR
□:
        (O13DIAX+O14DIAX)÷2
```

```
GIVE THE POSITION OF THE FIRST REFERENCE POINT
□:
      C12DIAX
GIVE THE POSITION OF THE SECOND REFERENCE POINT
□:
      O1DIAX

THE REFERENCE POINTS IN THE NEW COORDINATE SYSTEM
DONOR                   :      .00000     .00000    .00000
FIRST   REFERENCE POINT:     -.54950     .00000    .00000
SECOND REFERENCE POINT:    -2.76884    1.00075    .00000
```

POSITIONS OF THE PROTONS IN THE NEW COORDINATE SYSTEM

PROTON	X	Y	Z
1	−2.78772	3.01743	−.32657
2	−1.68193	2.77838	−2.48261
3	−1.83521	.64999	−3.53262
4	−3.33100	−.79644	−1.92342
5	−1.62553	−1.25373	−2.08086
6	−.57276	1.53007	.85535
7	.03873	2.04318	−.24352
8	−1.01736	3.28438	.83391
9	−.18551	1.49134	.13286
10	−1.13240	2.56462	1.31091
11	−.33488	3.24144	−.17594

```
IS THE LANTHANIDE POSITION GIVEN IN CARTESIAN (C)
OR SPHERICAL (S) COORDINATES?
C
ENTER THE LANTHANIDE POSITION
X  Y  Z
□:
      2.5  0  0
```

RELATIVE INDUCED SHIFTS

MOLAR FRACTION	1	2	3	4	5	ME
(FIRST FORM)						
.00	10.00	6.90	7.87	12.62	20.17	33.97
.10	10.00	7.09	8.19	14.34	20.73	31.21
.20	10.00	7.27	8.49	15.97	21.25	28.59
.30	10.00	7.44	8.78	17.51	21.74	26.12
.40	10.00	7.60	9.05	18.97	22.21	23.77
.50	10.00	7.76	9.31	20.36	22.65	21.54
.60	10.00	7.90	9.55	21.68	23.08	19.42
.70	10.00	8.04	9.79	22.93	23.48	17.40
.80	10.00	8.17	10.01	24.13	23.86	15.48
.90	10.00	8.30	10.22	25.27	24.23	13.64
1.00	10.00	8.42	10.43	26.36	24.58	11.89

Example 5: Simulation of Induced Shifts for Different Equilibria for DIEQ (HC-1) Conformation with Lanthanide Positioned at Axial or Equatorial

Carboethoxy Group, for PGM1

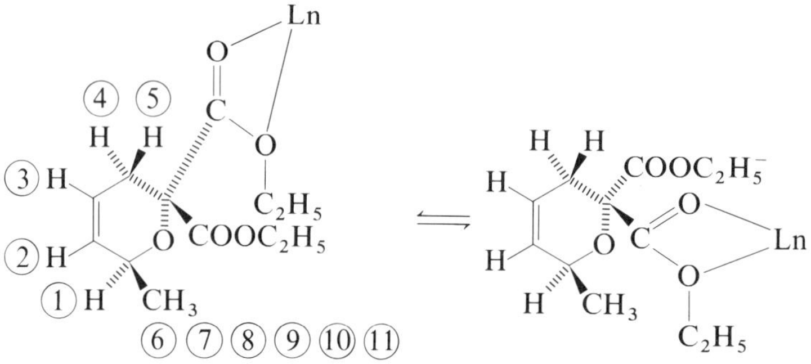

PGM1

DATA FOR THE FIRST FORM
HOW MANY PROTONS IN YOUR SYSTEM?
⎕:
11
GIVE THEIR POSITIONS (A N × 3 MATRIX)
⎕:
DIEQ
GIVE THE POSITION OF THE DONOR
⎕:
(O8DIEQ + O9DIEQ) ÷ 2
GIVE THE POSITION OF THE FIRST REFERENCE POINT
⎕:
C7DIEQ
GIVE THE POSITION OF THE SECOND REFERENCE POINT
⎕:
O1DIEQ

THE REFERENCE POINTS IN THE NEW COORDINATE SYSTEM

DONOR :	.00000	.00000	.00000
FIRST REFERENCE POINT:	−.54845	.00000	.00000
SECOND REFERENCE POINT:	−2.70323	1.14401	.00000

POSITIONS OF THE PROTONS IN THE NEW COORDINATE SYSTEM

PROTON	X	Y	Z
1	−4.35894	.45266	1.06002
2	−5.71111	.21953	−1.14376
3	−4.29914	−1.10615	−2.57180
4	−2.07022	−1.59395	−1.77522
5	−1.93619	.06470	−2.35484
6	−4.20555	2.89769	.95303
7	−5.85187	2.31656	.40000
8	−4.63233	2.84817	−.81147
9	−5.41823	2.54683	−.64437
10	−5.22778	2.58036	1.10293
11	−3.89848	3.04844	−.02266

```
IS THE LANTHANIDE POSITION GIVEN IN CARTESIAN (C)
OR SPHERICAL (S) COORDINATES?
C
ENTER THE LANTHANIDE POSITION
X  Y  Z
□:
      2.5  0  0

DATA FOR THE SECOND FORM
HOW MANY PROTONS IN YOUR SYSTEM?
□:
      11
GIVE THEIR POSITIONS (A N×3 MATRIX)
□:
      DIEQ
GIVE THE POSITION OF THE DONOR
□:
      (O13DIEQ+O14DIEQ)÷2
GIVE THE POSITION OF THE FIRST REFERENCE POINT
□:
      C12DIEQ
GIVE THE POSITION OF THE SECOND REFERENCE POINT
□:
      O1DIEQ


THE REFERENCE POINTS IN THE NEW COORDINATE SYSTEM
DONOR                  :      .00000     .00000    .00000
FIRST   REFERENCE POINT:     −.55378     .00000    .00000
SECOND REFERENCE POINT:    −2.44194    1.60168    .00000


POSITIONS OF THE PROTONS IN THE NEW COORDINATE SYSTEM
PROTON          X            Y           Z
----------------------------------------------
         1    −.68501     2.22368     .93235
         2   −1.75380     1.84491    3.26764
         3   −2.58114     −.41321    3.18514
         4   −2.47625    −1.55015    1.05514
         5   −3.88431     −.50462     .88409
         6   −2.18073     4.06609     .31470
         7   −1.70709     4.16656    2.08091
         8   −3.33905     3.56469    1.62044
         9   −2.77393     3.75528    2.23744
        10   −1.53820     4.34834    1.13665
        11   −3.10773     3.67641     .55513


IS THE LANTHANIDE POSITION GIVEN IN CARTESIAN (C)
OR SPHERICAL (S) COORDINATES?
C
ENTER THE LANTHANIDE POSITION
X  Y  Z
□:
      2.5  0  0
```

RELATIVE INDUCED SHIFTS

MOLAR FRACTION	1	2	3	4	5	ME
(FIRST FORM)						
.00	10.00	2.65	3.70	7.60	5.01	1.75
.10	10.00	2.83	3.88	8.16	5.77	1.98
.20	10.00	3.04	4.08	8.77	6.62	2.23
.30	10.00	3.27	4.30	9.44	7.55	2.51
.40	10.00	3.52	4.55	10.19	8.59	2.82
.50	10.00	3.81	4.82	11.03	9.75	3.16
.60	10.00	4.13	5.13	11.98	11.06	3.56
.70	10.00	4.50	5.49	13.05	12.55	4.00
.80	10.00	4.92	5.89	14.28	14.26	4.51
.90	10.00	5.40	6.36	15.71	16.24	5.10
1.00	10.00	5.97	6.91	17.38	18.55	5.79

Example 6: Ranges for the Induced Shifts if the Lanthanide Moves in a Cube, for PGM2

```
        PGM2

WHAT IS THE SIDE OF THE CUBE?
□:
        0.1
HOW MANY PROTONS IN YOUR SYSTEM?
□:
        11
GIVE THEIR POSITIONS (A N×3 MATRIX)
□:
        DIEQ
GIVE THE POSITION OF THE DONOR
□:
        (O8DIEQ+O9DIEQ+O13DIEQ+O14DIEQ)÷4
GIVE THE POSITION OF THE FIRST REFERENCE POINT
□:
        C6DIEQ
GIVE THE POSITION OF THE SECOND REFERENCE POINT
□:
        O1DIEQ


THE REFERENCE POINTS IN THE NEW COORDINATE SYSTEM
DONOR                  :     .00000     .00000    .00000
FIRST REFERENCE POINT  :   -1.24547     .00000    .00000
SECOND REFERENCE POINT:    -2.09049    1.14197    .00000
```

POSITIONS OF THE PROTONS IN THE NEW COORDINATE SYSTEM

PROTON	X	Y	Z
1	−1.97543	1.52369	2.04550
2	−4.07009	.04995	2.46959
3	−3.55483	−2.04102	1.39591
4	−1.53439	−2.13429	.07317
5	−2.67137	−1.30480	−.98714
6	−3.18373	3.24616	.78603
7	−4.18153	2.64111	2.19748
8	−4.56130	2.09154	.52676
9	−4.74156	2.07023	1.36547
10	−3.50000	3.22910	1.81937
11	−3.72975	2.67654	.11790

```
IS THE LANTHANIDE POSITION GIVEN IN CARTESIAN (C)
OR SPHERICAL (S) COORDINATES?
C
ENTER THE LANTHANIDE POSITION
X  Y  Z
□:
      2.5  0  0
```

PROTON	MINIMUM	INDUCED SHIFT	MAXIMUM
1	6.4027	6.5498	6.6719
2	3.2805	3.3342	3.3806
3	3.9343	3.9763	4.0167
4	10.0000	10.0000	10.0000
5	7.6125	7.6627	7.7243
6	2.7412	2.8169	2.8842

Listing of Programs

```
      ∇SIMULATION[□]∇

      ∇ SIMULATION;PROTONS;LANT;SHIFTS;N;I
[1]  A   ENTER THE POSITION OF THE PROTONS
[2]    PROTONS←CALLDATA
[3]  A   ENTER THE POSITION OF THE LANTHANIDE
[4]    LANT←CALLLANT
[5]  A   COMPUTE THE INDUCED SHIFTS
[6]    SHIFTS←LANT FACTOR PROTONS
[7]  A   SCALE THE SHIFTS SO THAT THE LARGEST ONE IS 10.0
[8]  A   IN ABSOLUTE VALUE
[9]    SHIFTS←SHIFTS×10÷SHIFTS[1↑ΨSHIFTS]
[10] A   PRINT OUT THE RESULTS
[11]   N←ρSHIFTS
[12]   I←1
```

```
[13]   3  1 ρ' '
[14]   '   PROTON INDUCED'
[15]   '     SHIFT'
[16] LOOP:(6 0 ⍕I), (10 5 ⍕SHIFTS[I])
[17]   →(N≥I←I+1)/LOOP
       ∇


       ∇PGM1[□]∇

       ∇  PGM1;P1;P2;L;S1;S2;SHIFT;I;X1
[1]  A    THIS PROGRAM COMPUTE THE DIPOLAR SHIFT FOR
[2]  A    A MOLECULE WITH EITHER TWO FORMS (EG. 1-C OR C-1)
[3]  A    OR TWO DONORS WITH SAME TYPE OF COMPLEXATION WITH
[4]  A    THE LANTHANIDE ION.
[5]    'DATA FOR THE FIRST FORM'
[6]    P1←CALLDATA
[7]    3 1 ρ' '
[8]    L1←CALLLANT
[9]    S1←L1 FACTOR P1
[10]   S1←ADJUST S1
[11]   PAGE←⍞
[12]   PAGE←⍞
[13]   'DATA FOR THE SECOND FORM'
[14]   P2←CALLDATA
[15]   3 1 ρ' '
[16]   L2←CALLLANT
[17]   S2←L2 FACTOR P2
[18]   S2←ADJUST S2
[19]   SHIFT←((1+ρS1),11)ρ0
[20]   I←1
[21]   X1←0
[22] LOOP:S←(X1×S1)+(1−X1)×S2
[23]   SHIFT[;I]←X1,S×10÷S[1]
[24]   X1←X1+0.1
[25]   →(11≥I←I+1)/LOOP
[26]   PAGE←⍞
[27]   PAGE←⍞
[28]   '                              RELATIVE INDUCED SHIFTS'
[29]   'MOLAR          ',45ρ'−'
[30]   'FRACTION        1          2          3          4          5          ME'
[31]   '--------      -----      -----      -----      -----      -----      -----'
[32]   '(FIRST FORM)'
[33]   8 2 ⍕(⍉SHIFT)
       ∇


       ∇PGM2[□]∇
       ∇  PGM2;I;N;PROTON;LANT;E1;E2;E3;M1;M2;M3;SHIFT;SHIFTS;EXTREMA;CONS
          T
[1]  A    THIS PROGRAM FIND AN ESTIMATE FOR THE RANGE OF RELATIVE
[2]  A    INDUCED SHIFTS FOR A PROTON IF THE LANTHANIDE IS
[3]  A    ALLOWED TO MOVE INSIDE A CUBE OF SIDE 'CONST' ANGSTROM.
[4]    'WHAT IS THE SIDE OF THE CUBE?'
[5]    CONST←0.5×□
[6]    PROTON←CALLDATA
[7]    3 1 ρ' '
[8]    LANT←CALLLANT
```

```
[9]    E1← 1 0 0
[10]   E2← 0 1 0
[11]   E3← 0 0 1
[12]   M1← −1 0 0
[13]   M2← 0 −1 0
[14]   M3← 0 0 −1
[15]   SHIFT←OFFSIDE 0,0,0
[16]   SHIFTS←((ρSHIFT),8)ρ0
[17]   SHIFTS[;1]←OFFSIDE E1+E2+E3
[18]   SHIFTS[;2]←OFFSIDE E1+E2+M3
[19]   SHIFTS[;3]←OFFSIDE E1+M2+E3
[20]   SHIFTS[;4]←OFFSIDE E1+M2+M3
[21]   SHIFTS[;5]←OFFSIDE M1+E2+E3
[22]   SHIFTS[;6]←OFFSIDE M1+E2+M3
[23]   SHIFTS[;7]←OFFSIDE M1+M2+E3
[24]   SHIFTS[;8]←OFFSIDE M1+M2+M3
[25]   I←1
[26]   EXTREMA←(2,(ρSHIFT))ρ0
[27]   N←ρSHIFT
[28] LOOP:EXTREMA[;I]←MAXMIN SHIFTS[I;]
[29]   →(N≥I←I+1)/LOOP
[30]   3 1 ρ' '
[31]   'PROTON          MINIMUM     INDUCED SHIFT        MAXIMUM'
[32]   '--------        ---------   --------------       ---------'
[33]   I←1
[34] LOOP2:(6 0 ⍕I), (16 4 ⍕(EXTREMA[2;I],SHIFT[I],EXTREMA[1;I]))
[35]   →(N≥I←I+1)/LOOP2
       ∇
```

```
       ∇ADJUST[⎕]∇

       ∇ S←ADJUST T
[1]  ⍝   THE LAST 6 'PROTONS' ARE THE METHYL. WE AVERAGE THEIR
[2]  ⍝   INDUCED SHIFTS.
[3]      S←(−6↓T),(+/−6↑T)÷6
[4]  ⍝   WE SCALE THE INDUCED SHIFTS SO THAT THE LARGEST ONE
[5]  ⍝   IS (+ OR −) 10.0 PPM.
[6]      S←S×10÷S[¯1↑⍋S]
       ∇
```

Subroutines

```
       ∇CALLDATA[⎕]∇

       ∇ PROTONS←CALLDATA;PROTONS;REF;NBOFPROTONS;I
[1]    REF← 3 3 ρ0
[2]    'HOW MANY PROTONS IN YOUR SYSTEM ?'
[3]    NBOFPROTONS←⎕
[4]    'GIVE THEIR POSITIONS (A N×3 MATRIX)'
[5]    PROTONS←(NBOFPROTONS,3)ρ⎕
[6]    'GIVE THE POSITION OF THE DONOR'
[7]    REF[3;]←⎕
[8]    'GIVE THE POSITION OF THE FIRST REFERENCE POINT'
[9]    REF[1;]←⎕
[10]   'GIVE THE POSITION OF THE SECOND REFERENCE POINT'
```

```
[11]  REF[2;]←□
[12]  PROTONS←POSITION PROTONS
[13]  3 1 ρ' '
[14]  'THE REFERENCE POINTS IN THE NEW COORDINATE SYSTEM'
[15]  ('DONOR                    :  '),(10 5  ⍕REF[3;])
[16]  ('FIRST    REFERENCE POINT:  '),(10 5  ⍕REF[1;])
[17]  ('SECOND REFERENCE POINT:  '),(10 5  ⍕REF[2;])
[18]  3 1 ρ' '
[19]  'POSITIONS OF THE PROTONS IN THE NEW COORDINATE SYSTEM'
[20]  'PROTON      X          Y          Z'
[21]  '------    ------    ------    ------'
[22]  I←1
[23] LOOP:(6 0  ⍕I),(10 5  ⍕PROTONS[I;])
[24]  →(NBOFPROTONS≥I←I+1)/LOOP
      ∇


      ∇CALLLANT[□]∇

      ∇LANT←CALLLANT;ANSWER
[1]   'IS THE LANTHANIDE POSITION GIVEN IN CARTESIAN (C)'
[2]   'OR SPHERICAL (S) COORDINATES ?'
[3]  A  CARTESIAN COORDINATE IS THE DEFAULT.
[4]   ANSWER←⍞
[5]   'ENTER THE LANTHANIDE POSITION'
[6]   →(ANSWER='S']/SPHERICAL
[7]   'X   Y   Z'
[8]   LANT←□
[9]   →JUMP
[10] SPHERICAL: 'RADIUS OMEGA PHI'
[11]  LANT←□
[12]  LANT←CONVERT LANT
[13]  'THE CARTESIAN COORDINATES OF THE LANTHANIDE ARE:'
[14]  '(X,Y,Z) =  ',(10 4   ⍕LANT)
[15] JUMP:
      ∇


      ∇CONVERT[□]∇

      ∇ LANTCART←CONVERT LANTSPHER;R;OMEGA;PHI;X;Y;Z;MULTIPLIER
[1]  A   MULTIPLIER TO CONVERT FROM DEGREE TO RADIANS
[2]    MULTIPLIER←(○1)÷180
[3]  A   ENTER THE COORDINATES
[4]    R←LANTSPHER[1]
[5]    OMEGA←LANTSPHER[2]×MULTIPLIER
[6]    PHI←LANTSPHER[3]×MULTIPLIER
[7]  A  X ← −R × COS(OMEGA)
[8]    X←−RX(2○OMEGA)
[9]  A  Y ← R × SIN(OMEGA) × COS(PHI)
[10]   Y←R×(1○OMEGA)×(2○PHI)
[11] A  Z ← R × SIN(OMEGA) × SIN(PHI)
[12]   Z←R×(1○OMEGA)×(1○PHI)
[13] A   RETURN THE CONVERTED POSITION
[14]  LANTCART←X,Y,Z
      ∇
```

```
     ∇POSITION[□]∇

     ∇ NEWP←POSITION OLDP;MAT
[1]  A   GLOBAL VARIABLE:REF
[2]  A   OLDP AND NEWP ARE THE COORDINATES OF THE PROTONS
[3]  A   IN THE ORIGINAL AND THE INTERMEDIATE COORDINATE
[4]  A   SYSTEMS, RESPECTIVELY
[5]  A
[6]  A   TRANSLATE THE SYSTEM SUCH THAT THE DONOR (REF[3;])
[7]  A   WILL BE ON THE ORIGIN
[8]    NEWP←OLDP-(ρOLDP)ρREF[3;]
[9]    REF←REF-(ρREF)ρREF[3;]
[10] A   COMPUTE THE ROTATION MATRIX
[11]   MAT←ROTATE REF
[12] A   ROTATE THE SYSTEM
[13]   NEWP←ØMAT+.×ØNEWP
[14]   REF←ØMAT+.×ØREF
[15] A   REF[1;] IS NOW ON THE X-AXIS AND
[16] A   REF[2;] IS IN THE XY-PLANE.
     ∇


     ∇FACTOR[□]∇

     ∇ SHIFTS←LANT FACTOR PROTONS;Q;R;D;L;COS
[1]  A   LANT IS THE CARTESIAN POSITION OF THE LANTHANIDE
[2]  A   PROTONS[I;] IS THE POSITION OF THE I-TH PROTON IN
[3]  A   THE NEW COORDINATE SYSTEM.
[4]  A   Q[I;] IS THE VECTOR LANT→PROTONS[I;]
[5]    Q←PROTONS-(ρPROTONS)ρLANT
[6]  A   R[I] IS THE DISTANCE LANT→PROTONS[I;]
[7]    R←(+/Q*2]*0.5
[8]  A   D[I;] IS THE DIRECTION COSINES OF THE VECTOR Q[I;]
[9]    D←ØQ÷Ø(ØρPROTONS)ρR
[10] A   L IS THE DIRECTION COSINES OF THE VECTOR DONOR→LANT
[11]   L←LANT÷(+/LANT*2)*0.5
[12] A   COS[I] IS THE COSINE OF THE ANGLE BETWEEN VECTORS D[I;] AND L
[13]   COS←L+.×D
[14] A   SHIFTS[I] IS THE INDUCED SHIFT FOR THE I-TH PROTON
[15]     SHIFTS←((3×COS*2)−1)÷R*3
     ∇


     ∇PLANE[□]∇

     ∇ LMN←PLANE Q;P1;P2;P3;ABC
[1]    P1←3↑Q
[2]    P2←3↑3↓Q
[3]    P3←6↓Q
[4]  A   GIVEN THAT THE THREE POINTS P1,P2, AND P3 ARE NOT
[5]  A   COLINEAR, THEY DEFINE A PLANE WHOSE NORMAL HAS
[6]  A   DIRECTION NUMBERS A,B,C AND DIRECTION COSINES L,M, AND N.
[7]    P2←P2−P1
[8]    P3←P3−P1
[9]    ABC←(P2[2 3 1]×P3[3 1 2])−(P2[3 1 2]×P3[2 3 1])
[10] LMN←ABC÷(+/ABC*2)*0.5
     ∇
```

```
      ∇ROTATE[□]∇

      ∇ MAT←ROTATE REF
[1]  A   MAT WILL BE THE ROTATION MATRIX (3×3) NEEDED
[2]  A   TO PLACE THE FIRST REFERENCE POINT ON THE
[3]  A   X-AXIS AND THE SECOND ONE IN THE XY-PLANE.
[4]     MAT←3 3 ρ0
[5]  A   MAT[1;] IS THE DIRECTION COSINES OF THE VECTOR
[6]  A   DONOR → REF[1;]
[7]     MAT[1;]←REF[1;]÷(+/REF[1;]*2)*0.5
[8]  A   MAT[2;] IS THE DIRECTION COSINES OF THE NORMAL
[9]  A   TO THE PLANE DEFINED BY THE DONOR, A POINT ON
[10] A   MAT[1;] AND REF[2;]
[11]    MAT[2;]←PLANE REF[3;],MAT[1;],REF[2;]
[12] A   MAT[3;] IS THE DIRECTION COSINES OF THE NORMAL
[13] A   TO THE PLANE DEFINED BY THE DONOR, A POINT ON
[14] R   MAT[1;] AND A POINT ON MAT[2;]
[15]    MAT[3;]←PLANE REF[3;],MAT[1;],MAT[2;]
[16] R   AS IT STANDS NOW, MAT WILL ROTATE THE FIRST
[17] A   REFERENCE POINT ON THE +X AXIS AND THE SECOND
[18] A   ONE IN THE XZ-PLANE.
[19]    MAT←(3 3 ρ−1,0,0,0,0,−1,0,−1,0)+.×MAT
[20] A   WITH THIS ADDED ROTATION, MAT WILL ROTATE THE
[21] A   FIRST REFERENCE POINT ON THE −X AXIS AND THE SECOND
[22] A   ONE IN THE +Y REGION OF THE XY-PLANE.
      ∇


      ∇MAXMIN[□]∇

      ∇ MM←MAXMIN V
[1]     T←[φV]
[2]     MM←(−1↑T),T[1]
      ∇


      ∇OFFSIDE[□]∇

      ∇ T←OFFSIDE V
[1]  A   GLOBAL: LANT,PROTON,CONST
[2]     T←(LANT+CONST×V) FACTOR PROTON
[3]     T←ADJUST T
[4]  A   WE HAVE COMPUTED THE INDUCED SHIFTS AS IF THE
[5]  A   LANTHANIDE WAS MOVED 'CONST' ANGSTROM IN THE
[6]  A   DIRECTION OF VECTOR V.
      ∇
```

Similar Programs

Since the advent of NMR–LSR studies, many computer programs have been written to solve such problems as discussed here. Unfortunately, most are no-name programs; further, their availability is somewhat in question because they have been written some time ago; preoccupations of the researchers may have shifted since. Nevertheless, the algorithms used can be retrieved or reconstructed from some of the researchers' papers.

Following is a list of the programs and their uses we were able to compile.

CHMSHIFT[3,72] finds the optimum geometry of the complex. The direction of the magnetic axis need not be colinear to the donor–Ln bond.

PDIGM[18,73] finds the optimum geometry of the complex.

OPLIS[74] finds the optimum geometry of the complex.

MAXI[53] finds the optimum geometry of the complex, by modifying the conformation (during computation) to obtain a better fit.

LANTHAMATIC[5] finds the optimum geometry of the complex.

Two of the programs without names are worth mentioning. In one, by ApSimon,[70] instead of searching for the lanthanide optimum position by looking at every grid position, this program uses Newton's method, so the search is more rapid. In another program, by Montando et al,[53,74] the ratio of two conformers is computed using the LSR technique.

The other no-name (one is inclined to use the word "generic" here) programs[8,22,27,30,37,65,75–83] in general use the same kind of search patterns for the lanthanide optimal position. A grid is defined with equal spacing in x, y, z; the induced shifts are computed at every point of the grid; then these shifts are compared with the experimental values. The point with the best correlation is declared the optimal lanthanide position. Sometimes the grid is defined in term of R, ϕ, and Ω, a spherical coordinate system the origin and orientation of which may be different in different programs.

The list of references given here is far from being exhaustive. However, it should provide an idea of what can be done by the use of LSR conformational methods, and how to do it.

The Choice of the Language Used: APL

Our programs have been written in APL. This choice has been made for several reasons. The language is a very compact one; a mathematical formula (even one dealing with matrices and vectors) usually can be translated into one APL statement; a third and every important reason is that interfaces between the main program and the subroutines are accomplished easily.

APL is one of the easiest languages for writing programs; without much exaggeration it can be compared to writing mathematical formulas directly.

The simplicity of the language is outstanding. Moreover, if a computer specialist is available to help an experienced person write a program, such a person can achieve independence in programming in a relatively short time, sooner than with other, less conversational languages.

The books by Gilman and Rose[84] and by LePage[85] constitute two quite satisfactory introductions to APL. Nevertheless, it may be a wise precaution to consult local computing services for updates in their versions of APL.

Acknowledgments

This work was supported by the Research Council of the Université de Moncton. We are indebted to Professors F. Söler and J. Sichel of our university for helpful comments. We would like to express our very special thanks to Dr. W. Brostow of Drexel University in Philadelphia for encouragement and friendly editorial criticism. One of us (K.J.) is particularly grateful to the Nuclear Research Centre (CEN) of Saclay (France) for the use of their facilities. Finally, we thank Mr. J. R. Boudreau for his excellent graphical service.

References

1. McConnell, H. M., Robertson, R. E.; *J. Chem. Phys.*, **1958**, *29*, 1361.
2. Jankowski, K. *Chem. Can.*, **1979**, *31*(5), 15.
3. Hawkes, G. E.; Leibfritz, D.; Roberts, D. W.; Roberts, J. D. *J. Am. Chem. Soc.*, **1973**, *95*, 1659.
4. Jack, A. *Acta Crystallogr.*, **1977**, *A33*(3), 497.
5. Rafaiski, A. J.; Barciszewski, J.; Karonski, M. *J. Mol. Struct.*, **1973**, *19*(1), 223.
6. Raber, D. J.; Janks, C. M.; Johnston, M. D.; Schwalke, M. A.; Shapiro, B. L.; Behelfer, G. L. *J. Org. Chem.*, **1981**, *46*, 2528.
7. Kosobutskii, V. A.; Dorofeeva, I. B. *Zh. Fiz. Khim.*, **1978**, *52*, 490.
8. Pirkle, W. H.; Rinaldi, P. L.; Simmons, K. A. *J. Magn. Reson.*, **1979**, *34*, 251.
9. Davis, R. E.; Willcott, R. M. *J. Am. Chem. Soc.*, **1972**, *94*, 1744.
10. Thompson, H. B. *J. Chem. Phys.*, **1970**, *52*, 3034.
11. Eyring, H. *Phys. Rev.*, **1932**, *39*, 746.

12. Hilderbrandt, R. L. *J. Chem. Phys.*, **1969**, *51*, 1654.
13. Thompson, H. B. *J. Chem. Phys.*, **1967**, *47*, 3407.
14. Nordlander, J. E. *J. Chem. Ed.*, **1973**, *50*(11), 743, and refs. quoted therein.
15. Briggs, J. M.; Moss, G. P.; Randall, E. W.; Sales, K. D. *J. Chem. Soc., Chem. Commun.*, **1972**, 1180.
16. Horrocks, W. DeW., Jr. *J. Am. Chem. Soc.*, **1974**, *96*, 3022.
17. Hawkes, G. E.; Marzin, C.; Johns, S. R.; Roberts, J. D. *J. Am. Chem. Soc.*, **1973**, *95*, 1661.
18. Willcott, M. R., III; Lenkinski, R. E.; Davis, R. E. *J. Am. Chem. Soc.*, **1972**, *94*, 1742.
19. Wing, R. M.; Early, T. A.; Uebel, J. J. *Tetrahedron Lett.*, **1972**, 4153.
20. Marinetti, T. D.; Snyder, G. H.; Sykes, B. D. *J. Am. Chem. Soc.*, **1975**, *97*, 6562.
21. Briggs, J. M.; Hart, F. A.; Moss, G. P.; Randall, E. W.; Sales, K. D.; Staniforth, M. L. In "Nuclear Magnetic Resonance Shift Reagents"; Sievers, R. E., Ed.; Academic Press: New York, **1973**; p. 197.
22. Wing, R. M.; Uebel, J. J.; Anderson, K. K. *J. Am. Chem. Soc.*, **1973**, *95*, 6046.
23. Dyer, D. S.; Cunningham, J. A.; Brooks, J. J.; Sievers, R. E.; Rondeau R. E. In "Nuclear Magnetic Resonance Shift Reagents"; Sievers, R. E., Ed.; Academic Press: New York, **1973**.
24. DeBoer, J. W. M.; Sakkers, P. J. D.; Hilbers, C. W.; DeBoer, E. *J. Magn. Reson.* **1977**, *25*, 455.
25. Briggs, J. M.; Hart, F. A.; Moss, G. P.; Randall, E. W.; Sales, K. D.; Staniforth, M. L. *Chem. Commun.*, **1951**, 364.
26. Briggs, J. M. Personal communication.
27. Richardson, M. F.; Rothstein, S. M.; Li, W.-K. *J. Magn. Reson.*, **1979**, *36*, 69.
28. Hamilton, W. C.; *Acta Crystallogr.*, **1965**, *18*, 502.
29. Horrocks, W. DeW. Jr.; Sipe, J. P., III. *Science*, **1972**, *177*, 994.
30. Briggs, J. M. Program described in ref. 15.
31. Farid, S.; Ateya, A.; Magio, M. *Chem. Commun.*, **1971**, 1285.
32. Demarco, P. V. Personal communication.
33. Ochiai, M.; Mizuta, E.; Aki, O.; Morimoto, A.; Okada, T. *Tetrahedron Lett.*, **1972**, 3245.
34. Siddal, T. H., III. *Chem. Commun.*, **1971**, 452.
35. Shapiro, B. L.; Heubucek, J. R.; Sullivan, G. R.; Johnson, L. F. *J. Am. Chem. Soc.*, **1971**, *93*, 3281.
36. Ammon, H. L.; Mazzochi, P. H.; Kopecky, W. J.; Jr.; Tamburin, H. J., Watt, P. H.; Jr., *J. Am. Chem. Soc.*, **1973**, *95*, 1968.
37. Briggs, J. M. Personal communication.
38. Honeyburne, C. L. *Tetrahedron Lett.*, **1972**, 1095.
39. Newman, R. H. *Tetrahedron*, **1974**, *30*, 969.
40. Huber, H. *Tetrahedron Lett.*, **1972**, 3559.
41. Raber, D. J.; Janks, C. M.; Johnston, M. D. Jr.; Raber, N. K. *J. Am. Chem. Soc.*, **1980**, *102*, 6591.
42. Barry, C. D.; Dobson, C. M.; Sweigart, D. A.; Ford, L. E.; Williams, R. J. P. In "Nuclear Magnetic Resonance Shift Reagents"; Sievers, R. E., Ed.; Academic Press: New York, 1973; p. 173.
43. Davis, R. E.; Willcott, M. R.; Lenkinski, R. E.; Doering, W. von E.; Birladeanu, L. *J. Am. Chem. Soc.*, **1973**, *95*, 6846.
44. Johnston, M. D.; Raber, D. J.; DeGennaro, N. K.; D'Angelo, A.; Perry, J. W. *J. Am. Chem. Soc.*, **1976**, *98*, 6042.
45. Raber, D. J.; Johnston, M. D.; Schwalke, M. A. *J. Am. Chem. Soc.*, **1977**, *99*, 7671.
46. Hawkes, G. E.; Marzin, C.; Leibfritz, D.; Johns, S. R.; Herwig, K.; Cooper, R. A.; Roberts, D. W.; Roberts, J. D. In "Nuclear Magnetic Resonance Shift Reagents"; Sievers, R. E., Ed.; Academic Press: New York, 1973; p. 129.

47. Montando, G., Caccamese, S.; Librando, V.; Maravigna, P. *Tetrahedron*, **1973**, *29*, 3915.
48. Sokalsi, W. A.; *Comput. Chem.*, **1980**, *4*(3, 4), 165 (Engl.).
49. Dillen, J.; Geise, H. J. *Comput. Chem.*, **1980**, *4*(3, 4), 113 (Engl.).
50. Lopata, A.; Kiss, A. I. *Comput. Chem.*, **1979**, *3*(2-4), 107 (Engl.).
51. Bundura, A. V.; Novogelev, N. P. *Zh. Strukt, Khim.*, **1978**, *19*(2), 357 (Russian).
52. Crippen, G. M.; Havel, T. F. *Acta Crystallogr.*, **1978**, *A34*(2), 282 (Engl.).
53. Sullivan, G. R. *J. Am. Chem. Soc.*, **1976**, *98*, 7162.
54. Corey, E. J. *J. Am. Chem. Soc.*, **1955**, *77*, 2505.
55. Hinckley, C. C.; Brumley, W. C. *J. Am. Chem. Soc.*, **1976**, *98*, 1331.
56. Levine, B. A.; Thornton, J. M.; Williams, J. P. *J. Chem. Soc., Chem. Commun.*, **1974**, *16*, 669.
57. Cockrill, A. F.; Davies, G. L. O.; Harden, R. C.; Rackham, D. M. *Chem. Rev.*, **1973**, *73*, 553.
58. Mohyla, I.; Ksandr, Z.; Hazek, M.; Vodicka, L. *Coll. Czech. Chem. Commun.*, **1974**, *39*, 2935.
59. Briggs, J.; Hart, F. A.; Moss, G. P. *J. Chem. Soc., Chem. Commun.*, **1970**, 1506.
60. Schneider, H. J.; Weigand, E. F. *Tetrahedron*, **1975**, *31*, 2125.
61. Welti, D. H.; Linder, M.; Ernst, R. R. *J. Am. Chem. Soc.*, **1978**, *100*, 403.
62. Heigl, T.; Micklow, G. K. *Tetrahedron Lett.*, **1973**, 649.
63. Finocciaro, P.; Recca, A. Maravigna, P. Montando, G. *Tetrahedron*, **1974**, *30*, 4159.
64. Chadwick, D. J. *Tetrahedron Lett.*, **1974**, 1375.
65. Chadwick, D. J.; Williams, D. H. *J. Chem. Soc., Chem. Commun.*, **1974**, *12*, 128.
66. Ammon, H. L.; Mazzochi, P. H.; Colicelli, E. J. *Org. Magn. Reson.*, **1978**, *11*, 1.
67. Lenkinski, R. E.; Reuben, J. *J. Am. Chem. Soc.*, **1976**, *98*, 4065.
68. Gray, A. I.; Waigh, R. D.; Waterman, P. G. *J. Chem. Soc., Chem. Commun.*, **1974**, 632.
69. Caple, R.; Harris, D. H.; Kuo, S. C. *J. Org. Chem.*, **1973**, *38*, 381.
70. Jankowski, K.; Couturier, J. E. *J. Org. Chem.*, **1972**, *37*, 3997.
71. ApSimon, J. W.; Beierbeck, H. *Tetrahedron Lett.*, **1973**, 581.
72. Roberts, J. D.; Hawkes, G. E.; Husar, J.; Robert, A. W.; Roberts, D. W. *Tetrahedron*, **1974**, *30*, 1833.
73. Davis, R. E.; Willcott, M. R. In "Nuclear Magnetic Resonance Shift Reagents"; Sievers, R. E., Ed.; Academic Press: New York, 1973; p. 143.
74. Kordova, I.; Fomichev, A. F.; Zvolinskii, V. P. *Zh. Struckt. Khim.*, **1971**, *22*, 27.
75. Morallee, K. G.; Nieboer, E.; Rossotti, F. J. C.; Williams, R. J. P.; Xavier, A. V. *Chem. Commun.*, **1970**, 1132.
76. Demarco, P. V.; Cerimele, B. J.; Crane, R. W.; Thakkar, A. L. *Tetrahedron Lett.*, **1972**, 3539.
77. Montando, G.; Finocchiaro, P. *J. Org. Chem.*, **1972**, *37*, 3434.
78. Heigl, T.; Mucklow, G. C.; *Tetrahedron Lett.*, **1973**, 649.
79. Montando, G.; Librando, V.; Caccamese, S.; Maravigna, P. *J. Am. Chem. Soc.*, **1973**, *95*, 6365.
80. A program by Dr. C. C. Hinckley described in ref. 8.
81. Armitage, I. M.; Hall, L. D.; Marshall, A. G.; Werebelow, L. G. *J. Am. Chem. Soc.*, **1973**, *95*, 1437.
82. Sanders, J. K. M.; Williams, D. H. *J. Am. Chem. Soc.*, **1971**, *93*, 641.
83. Hubert, H.; Pascual, C. *Helv. Chim. Acta.*, **1971**, *54*, 913.
84. Gilman, L.; Rose, A. J. "APL: An Interactive Approach"; J. Wiley & Sons: New York, 1974; 378 pp.
85. LePage, W. R. "Applied APL Programming"; Prentice-Hall: Englewood Cliffs, N.J., 1978; 308 pp.

Additional References

Tomisawa, H.; Hongo, H.; Chiasson, J. B.; Jankowski, K. *Heterocycles*, **1982**, *19*, 1257.
Chiasson, J. B.; Diaz, E., and Jankowski, K. *Proc. Indian Nat. Sci. Acad.*, **1983**, *49*, 193.
Jankowski, K.; Israeli, J.; Chiasson, J. B.; Rabczenko, A. *Heterocycles*, **1982**, *19*, 1215.
Jankowski, K. *Bull. Sci.*, **1978**, *26*, 737.
Jankowski, K. *Rev. Latin. Quim.*, **1978**, *9*, 107.
Jankowski, K. *J. Org. Chem.*, **1976**, *41*, 4103.
Jankowski, K.; Rabczenko, A. *Bull. Acad. Pol. Sci., Sec. Chem.*, **1976**, *24*, 453.
Jankowski, K.; Diaz, E.; Yuste, F.; and Walls, F. *J. Org. Chem.*, **1976**, *41*, 4103.
Jankowski, K.; Rabczenko, A. *J. Org. Chem.*, **1975**, *40*, 960.
Jankowski, K.; Israeli, J. *Bull. Acad. Pol. Sci.*, **1974**, *22*, 3.
Jankowski, K.; Pelletier, O.; Tower, R. *Bull. Acad. Pol. Sci.*, **1974**, *22*, 867.

3

THE NATURE OF THE LSR–SUBSTRATE COMPLEX

Douglas J. Raber

Department of Chemistry,
University of South Florida,
Tampa, Florida 33620

Introduction

Over the last decade the status of lanthanide shift reagents has undergone several important changes. Following Hinckley's report in 1969,[1] it appeared that shift reagents might provide the answer to virtually any difficult problem in NMR spectroscopy. It was soon recognized, however, that a substantial number of variables was involved in the interactions of shift reagents with organic molecules. As a consequence, the technique did not offer the simple solution that many had hoped for, and interest in the field declined substantially. In the last several years a number of careful studies have been carried out that have helped to resolve the difficulties of using and interpreting shift reagent data. These advances have now brought us to a stage where lanthanide shift reagents can be used to great advantage as long as appropriate techniques are employed.

In this chapter the available evidence on the formation of complexes between organic substrates and lanthanide shift reagents is reviewed. A critical assessment of methods for experimental measurement of lanthanide induced shifts is provided and experimental techniques for obtaining the optimum amount of structural information are suggested.

Morrill (ed): LANTHANIDE SHIFT REAGENTS IN STEREOCHEMICAL ANALYSIS

Many of the terms and abbreviations used in the field of lanthanide shift reagents are somewhat arbitrary and differ from one laboratory to the next. In order to minimize any possible confusion from this source, the following list summarizes the symbols and abbreviations used here.

L	a lanthanide shift reagent
S	an organic substrate
L_O	total concentration of lanthanide in both complexed and uncomplexed forms
S_O	total concentration of substrate in both complexed and uncomplexed forms
δ_O	the observed chemical shift of a nucleus in the absence of any shift reagent
δ_{obs}	the observed chemical shift of a nucleus in the presence of shift reagent
LIS	the lanthanide-induced shift of a nucleus, ie, the difference between δ_O and δ_{obs}.
Bound shift	the LIS of a nucleus in a specific complex between substrate and shift reagent
Ln	general abbreviation for any of the lanthanide elements
$Ln(dpm)_3$	the *tris*-dipivaloylmethanato derivative of a lanthanide; also $Ln(thd)_3$ for the ligand 2,2,6,6-tetramethylheptanedionato.
$Ln(fod)_3$	the *tris*-6,6,7,7,8,8,8-heptafluoro-2,2-dimethyloctanedionato derivative of a lanthanide

Stoichiometry of the Complex

Shift Reagent Equilibria

Although lanthanide-induced shifts frequently result in useful simplifications of NMR spectra, the greatest potential for shift reagents lies in the specific relationship between the magnitude of the LIS and the molecular structure of the lanthanide–substrate complex.[2–8] Unfortunately, the complexation equilibria are very rapid, and time-average spectra are nearly always obtained. This means that the observed chemical shifts are the weighted averages for the several species present in solution, and even in the simplest situation the substrate is present as a mixture of both the bound and unbound species. Frequently, more than one complex is involved in the equilibria, and this complicates the situation further, because rigorous interpretation of shift reagent experiments requires that the experimental LIS correspond to a single species rather than to some undefined mixture. If the maximum information about the molecular structure of a substrate is to be obtained from lanthanide-induced shifts, it is imperative to have a full understanding of the various equilibria that are involved.

Much of the early work with shift reagents was based on the assumption that only a single complex with 1:1 stoichiometry is formed in solution. Within several years, however, several groups presented evidence showing that more than one complex could be formed in solution.[9–14] x-Ray crystallographic structures also were reported, which demonstrated 7-coordinate,[15–20] as well as 8-coordinate[21–27] (and higher[28,29]), structures for lanthanides. In addition, the direct observation of a 2:1 complex at low temperature was reported.[30] Clearly the observed NMR spectra in the presence of shift reagents could not be interpreted properly unless it were known precisely which of the possible complexes had been formed.

The first indications that complexes of different stoichiometry could be formed in shift reagent experiments were reported in 1972.[9,10] Shortly thereafter, Shapiro and Johnston[11] presented a full mathematical analysis of lanthanide-induced shifts in terms of the equilibria shown in Figure 3-1. Their analysis included complexes of both 1:1 and 1:2 stoichiometry. Subsequent work by Reuben[14] and by Shapiro's group[31] provided strong support for this scheme as an acceptable model for shift reagent equilibria.

$$\mathrm{L + S \overset{K_1}{\rightleftharpoons} LS}$$

$$\mathrm{LS + S \overset{K_2}{\rightleftharpoons} LS_2}$$

Figure 3-1. Interaction of lanthanide shift reagent (L) with an organic substrate (S). Both 1:1 and 1:2 complexes can be formed.

One feature of shift reagent equilibria that is not included in Figure 3-1 is self-association of the shift reagent. This occurs when one or more of the chelate oxygens acts as a bridging ligand, coordinating simultaneously to both lanthanides. Both $Pr(dpm)_3$[15] and the hydrate of $Pr(fod)_3$[21] had been shown by x-ray crystallography to exist as dimers in the solid state. Moreover, several experiments had suggested that dimerization might also be important in solution.[32–36] DeBoer[36] found that the NMR data obtained with methyl butyl ether and $Pr(fod)_3$ suggested dimerization of the shift reagent, but the effect on the LIS was not large. Both Shapiro and his co-workers,[31] as well as Reuben[14] and

Hajeck,[37] have concluded that perturbations from dimerization of the shift reagent are small and generally can be ignored. The equilibria of primary interest for structural work are those involving the *substrate*, so dimerization of the shift reagent is unlikely to affect these to a great extent. Consequently, the equations shown in Figure 3-1 generally have been accepted as representing a good approximation of shift reagent equilibria in solution.

Most of the evidence for formation of both 1:1 and 1:2 complexes was obtained with fod reagents, and the question has remained whether or not the conclusions can be extended to dpm reagents as well. The latter are weaker Lewis acids,[2–8] and formation of the 1:2 complex should occur to a lesser extent. Indeed, it often has been assumed that only the 1:1 complex is formed with such shift reagents as $Eu(dpm)_3$. We reported previously that the presence of two different complexes could be detected from the behavior of the relative LISs obtained using $Eu(fod)_3$ with several adamantane derivatives, **1**–**3**.[38]

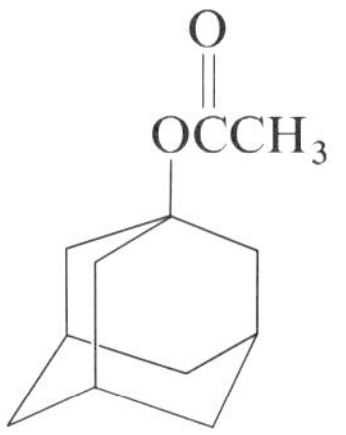

$OC(=O)CH_3$ $CH_2C(=O)OC_2H_5$ $CH_2C(OH)(CH_3)_2$

1 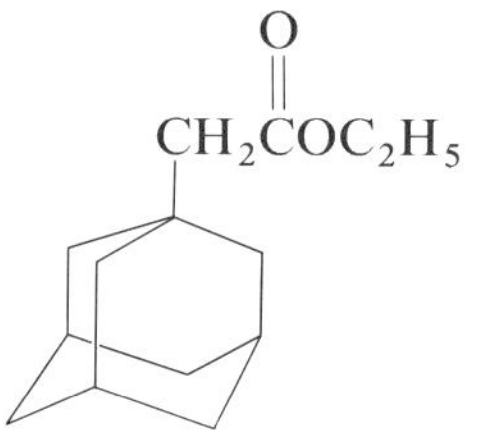   **2** 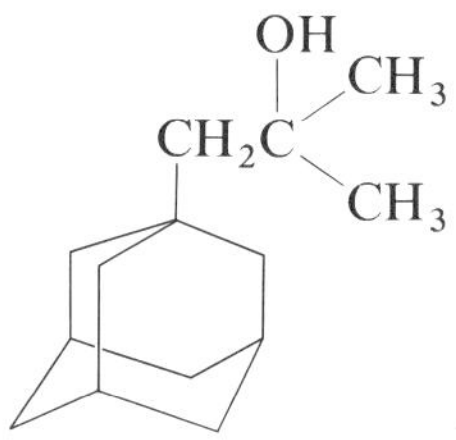   **3**

We repeated these experiments with $Eu(dpm)_3$ but were unable to find evidence for 1:2 complex formation.[39] This may partly reflect the lower solubility of $Eu(dpm)_3$, but the results nevertheless support the idea that 1:2 complex formation is less important for dpm derivatives. This does not mean that 1:2 complex formation is absent with dpm reagents, however. When dpm reagents are used with substrates that are relatively strong Lewis bases, the 1:2 adducts again form quite readily. This is demonstrated by the isolation of crystalline 1:2 complexes with pyridine,[24] 4-methylpyridine,[25] and dimethylformamide.[40] Similarly, Cramer[41] reported the direct observation in solution of such a complex between $Eu(dpm)_3$ and 4-methylpyridine by low-temperature NMR spectroscopy. Only in the case of substrates that are relatively weak Lewis bases (eg, ketones and nitriles[42–44]) does it appear that 1:2 adducts with dpm reagents may be formed to such a small extent that they can be neglected.

Direct Observation of Bound Shifts

Ideally, one would like to observe directly the NMR spectrum of a specific lanthanide–substrate complex, but this is usually precluded by fast ligand exchange. Only in a few instances has it been possible to obtain shifted spectra under the conditions of slow exchange. The earliest example of slow exchange was reported by Evans and Wyatt in 1972.[30] These authors investigated the interaction of deuterated $Eu(fod)_3$ with dimethylsulfoxide (DMSO) as a function of temperature, using a lanthanide–substrate concentration ratio of 0.3. At ambient temperature the DMSO gave rise to a single sharp peak at about 4.5 ppm, and this peak showed substantial broadening at −45°C. When the temperature was further lowered to −80°C, two separate peaks were observed. One peak was at about 2.7 ppm, corresponding to uncomplexed DMSO, and the other peak appeared at about 6.2 ppm. Integration of the peaks showed that the complex contained two substrate molecules, as would be expected at low lanthanide–substrate ratios where there is excess substrate. This initial report was confirmed by Reuben,[14] and Evans and Wyatt reported a more detailed study in 1974.[45] Their full study showed that slow exchange could be observed for a variety of $Ln(fod)_3$ reagents at low temperature. They were able to see discrete signals for both bound and free substrate with hexamethylphosphoramide (**5**), triethylamine (**6**), and tetramethylurea (**7**), as well as with dimethylsulfoxide (**4**).

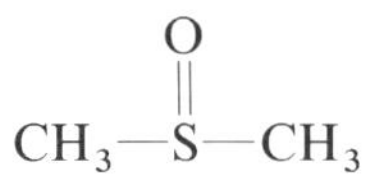

CH_3—S(=O)—CH_3

4

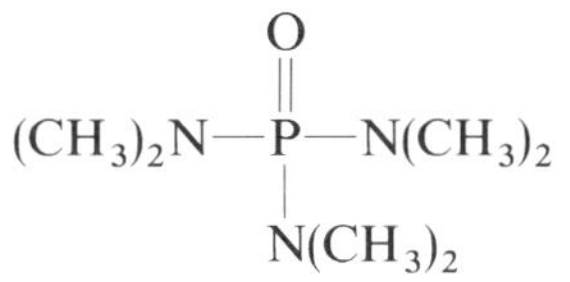

$(CH_3)_2N$—P(=O)($N(CH_3)_2$)—$N(CH_3)_2$

5

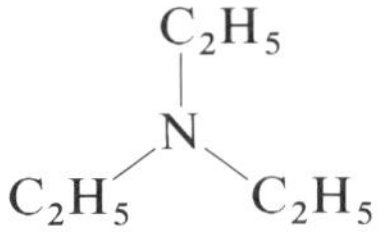

$N(C_2H_5)_3$

6

$(CH_3)_2N$—C(=O)—$N(CH_3)_2$

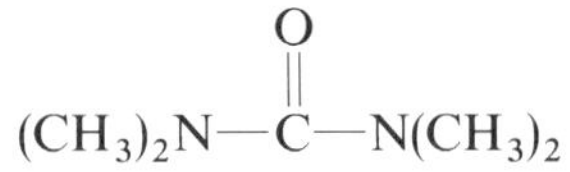

7

All of these compounds are relatively strong Lewis bases, and this is probably an important factor why slow exchange occurs. When $Eu(fod)_3$ was replaced by $Eu(dpm)_3$ (a weaker Lewis acid), it was not possible to obtain slow-exchange spectra, even at −100°C.

Integration of the low-temperature spectra showed that the 1:2 complex was observed in some cases, but in other instances the complex present in solution clearly had 1:1 stoichiometry. Although there was not sufficient information to define the conditions that favored one complex over the other, Evans and Wyatt unambiguously demonstrated that both complexes were important in shift reagent equilibria. Cramer and Dubois[41,46,47] found that slow-exchange spectra could be obtained for $Eu(dpm)_3$ with either 3-methylpyridine (**8**) or 4-methylpyridine (**9**) at $-110°C$ in carbon disulfide, and similar results were found with pyridine (**10**). Their experimental samples employed an excess of substrate, and, again, the 1:2 complex was observed.

More recently, Bovee and co-workers[48,49] were able to observe the 1:1 complex of quinuclidine (**11**) with $Yb(fod)_3$.

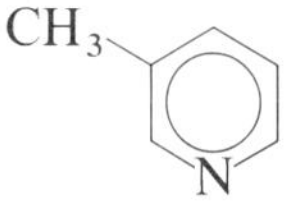

8 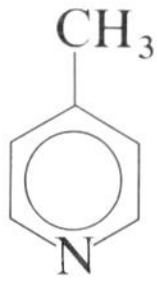 **9**  **10** 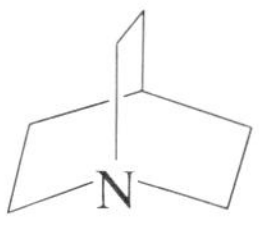 **11**

Several other examples of slow exchange have been reported for difunctional substrates that can chelate to a shift reagent. Grotens and co-workers[50] were able to observe directly the spectra of the complex between $Eu(fod)_3$ and 1,2-dimethoxyethane (**12**) at $-60°C$. They obtained similar results for low-temperature spectra of other polyethyleneglycol ethers (**13**).

$CH_3OCH_2CH_2OCH_3$ $\qquad$ $CH_3O(CH_2CH_2O)_nCH_3$

12 **13**

Lindoy and co-workers have found that several transition-metal acetylacetonates exhibit slow exchange with $Eu(fod)_3$, even at ambient temperature.[51,52] For example, at 32°C in deuteriochloroform solution the individual spectra of chromium complex **14** and its (deuterated) $Eu(fod)_3$ adduct can be observed simultaneously.[51]

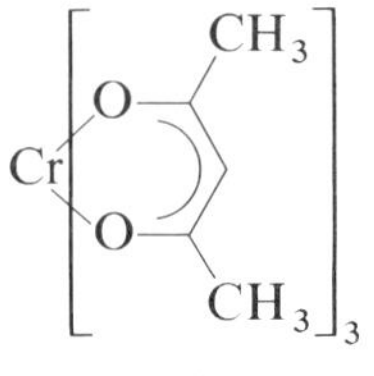

14

In principle, the direct observation of lanthanide–substrate complexes under slow-exchange conditions provides the most reliable structural information. Unfortunately, several factors prevent this in actual practice. First, slow exchange has been found in only a very limited number of examples. The pattern for these few cases suggests that a large association constant is necessary, and this in turn means that many (if not most) of the organic molecules of interest cannot be studied by this technique.[52a] A second difficulty arises from the complexity of low-temperature spectra. In addition to the presence of signals corresponding to both free and complexed substrate, other time-dependent processes can be resolved at low temperatures as well. For example, interconversion of substrate conformers also may be slow on the NMR time scale at the temperature necessary for slow ligand exchange. This would lead to highly complicated spectra that could be difficult to interpret. Still a third problem with direct observation of lanthanide–substrate complexes under slow-exchange conditions involves the temperature dependence of the LIS. The difference in chemical shift for a nucleus in the free and complexed substrate varies inversely with temperature. Moreover, the mathematical form of the temperature dependence is expected to be different for LISs that originate from contact and pseudocontact mechanisms.[53–59] The coalescence temperature is unlikely to be the same for different complexes, and this causes data to be obtained at different temperatures. Consequently, it would not be possible to correlate data for these various compounds unless different nuclei in a complex exhibited different coalescence temperatures. In that event data could be obtained over a range of temperatures and extrapolated to a common temperature. Although this may be possible in some cases, such procedures have not yet been reported.

Despite the clear advantages of directly observing induced shifts under slow-exchange conditions, this is not usually feasible. It therefore appears that extraction of bound shifts from time-averaged spectra provides the only general method for obtaining reliable and accurate LIS data.

Algorithms for Extracting Bound Shifts from Solution Data

Bound Shifts

In order to utilize lanthanide induced shifts for any sort of mathematical assessment of a proposed molecular structure, it is first necessary to obtain a numerical value for the LIS of each nucleus that is to be used in the evaluation. The procedure by which these numerical values are obtained is important, because the observed LISs typically reflect contributions from more than one species. Both 1:1 and 1:2 complexes commonly are formed, particularly with $(fod)_3$ reagents, and any attempt to analyze molecular structure on the basis of data from an undefined mixture is invalid.

The importance of utilizing bound shifts can be seen from inspection of Figure 3-2, which shows the relative LIS of 1-adamantyl acetate (**1**) as a function of $Eu(fod)_3$ concentration.[38] The hydrogens of the acetoxy methyl were selected for reference and therefore have a relative LIS of unity. The LISs of the remaining hydrogens are relatively independent of the $Eu(fod)_3$ concentration, except for those of the six equivalent methylene hydrogens (labeled B in the figure). The relative LIS of these hydrogens is less than 0.5 at low $Eu(fod)_3$ concentration (where the presence of excess substrate favors formation of the 1:2 complex). At higher concentrations of shift reagent, however, the relative LIS for the B hydrogens increases to 0.7. The latter conditions favor formation of the 1:1 adduct, so it is clear that quite different relative LIS can be observed

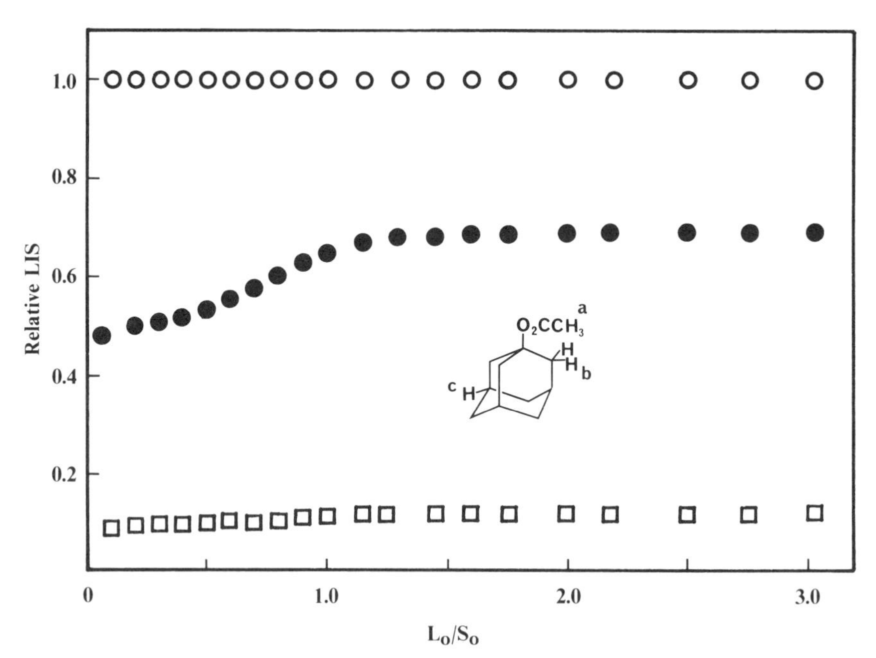

Figure 3-2. Relative LISs for 1-adamantyl acetate (**1**). Values are relative to those for the methyl group of the acetate (a, open circles). The six equivalent methylene hydrogens are designated type b (filled circles), and the three equivalent bridgehead hydrogens are designated type c (squares).

for the two complexes. Similar behavior has been reported for the LIS of quinoline by other groups[9,10]

Several strategies have been employed for determining bound shifts (ie, the induced shifts that correspond to formation of a specific complex). These range from complete analysis of all the data according to the scheme shown in Figure 3-1 to techniques that are at best approximate. In this section these different methodologies are reviewed and a critical assessment of their advantages and disadvantages is presented.[60]

The magnitude of a lanthanide-induced shift is defined according to eq. (3-1):

$$\mathrm{LIS} = \delta_{\mathrm{obs}} - \delta_{\mathrm{O}} \tag{3-1}$$

If both 1:1 and 1:2 complexes are present, then the observed LIS is equal to the sum of the contributions from each complex. Each complex contributes to the total shift by a quantity equal to the fraction of substrate in the specific complex multiplied by the bound shift of that complex:

$$\mathrm{LIS} = \frac{[\mathrm{LS}]}{\mathrm{S_O}}\Delta_1 + \frac{2[\mathrm{LS_2}]}{\mathrm{S_O}}\Delta_2 \tag{3-2}$$

Because fast exchange normally prevents direct measurement of the actual concentrations of the 1:1 and 1:2 complexes, the bound shifts must be obtained by further analysis of eq. (3-2). Although an explicit solution of eqs. (3-2)–(3-4) for the bound shifts is not possible, algebraic manipulation does permit simplification under certain conditions. Even when simplification is not appropriate, it is possible to determine both the bound shifts and the equilibrium constants by an iterative computer search for the values that yield "predicted" LISs in best agreement with experimental data.

Four methods have been employed frequently to determine bound shifts from observed LIS; here, these are designated as the two-step method, the gradient method, the reciprocal method, and the equimolar method. Only the first of these strategies fully takes into account the two equilibria of Figure 3-1 and eqs. (3-2)–(3-4). As a consequence, the other approaches can afford only approximate values for the bound shifts, and it is important to determine the validity of the approximations. Are the approximate values sufficiently close to the correct bound shifts and equilibrium constants that they justify the experimental simplifications, or do the experimental simplifications result in numerical data that are unreliable?

The shift reagent equilibria shown in Figure 3-1 can be described mathematically according to the two equilibrium constants, K_1 and K_2, described mathematically as:

$$K_1 = \frac{[LS]}{[L][S]} = \frac{[LS]}{(L_O - [LS] - [LS_2])(S_O - [LS] - 2[LS_2])} \tag{3-3}$$

$$K_2 = \frac{[LS_2]}{[LS][S]} = \frac{[LS_2]}{[LS](S_O - [LS] - 2[LS_2])} \tag{3-4}$$

Of all the terms in these equations only the observed LIS and the total concentrations (ie, corresponding to the total amounts, both free and complexed) of L and S are known. This leaves two bound shifts, two association constants, and three concentration terms, which are all unknown and cannot be measured directly. The number of unknown quantities is therefore greater than the number of independent equations relating them. Even if algebraic manipulation of eqs. (3-2)–(3-4) were to eliminate the concentration terms for the three complexes, four unknowns (two bound shifts and two association constants) would remain. Clearly, a direct and explicit solution of eqs. (3-2)–(3-4) is not possible.

Two alternatives therefore remain for determination of the bound shifts (and association constants). One approach that has seen extensive use is based on approximations that result in algebraic simplification of eqs. (3-2)–(3-4). The approximations can cause some of the terms in eqs. (3-2)–(3-4) to vanish, and direct analysis of the observed LIS then yields the desired bound shifts and/or association constants. A major difficulty has been the absence of reliable information for assessing the validity of the approximations that are used. Only recently has a detailed evaluation of this problem been carried out.[61] As shown in the next section, "Gradient Method", substantial errors may be present in values that are obtained by such methods. The second strategy for extracting bound shifts and association constants from the experimental data employs no assumptions (other than the validity of the model used—in this case the scheme shown in Figure 3-1). By means of an iterative fitting procedure, it is possible to determine those values of the bound shifts and association constants that best reproduce the experimental LIS when the latter are "predicted" with eqs. (3-2)–(3-4). Although the rigor of this approach argues in its favor,[61,62] it suffers from the inconvenience that experimental data must be obtained over a wide range of L:S ratios.

Gradient Method

The gradient method is based on a model that assumes the formation of only the 1:1 complex (ie, that $K_2 = 0$). The technique was developed prior to the time that 1:2 complex formation was established,[10,63,64] but it has been used extensively up to the present time. When the observed LIS for a nucleus is plotted against the ratio of shift reagent to substrate, the data points frequently describe a straight line (at least at low L:S ratios). The slope of this line is called the shift gradient. The extrapolation of LIS data to equimolar concentrations of shift reagent and substrate[64] constitutes an equivalent procedure. The gradient method probably has been used more frequently than any other procedure to determine LIS. The popularity of the method probably reflects its early introduction by workers in the field,[10,63,64] but it is also experimentally straightforward, and this has certainly provided an impetus for its continued use as well. Starting with a solution of substrate in the appropriate solvent, additions of small amounts of shift reagent provide the necessary solutions for spectroscopic observation. The small increments of shift reagent may be added as a solid (and the quantity can be recorded as a weight) or they may be added as a concentrated solution (thereby permitting measurement of the volume using a microliter syringe). In either case the change in total volume is considered typically to be negligible, so the total concentrations of substrate and shift reagent are easily calculated.

If the assumption of only 1:1 complex formation is valid, then algebraic manipulation of eqs. (3-2) and (3-3) permits the observed LIS to be described:[61]

$$\mathrm{LIS} = \left(\frac{\mathrm{L_O}}{\mathrm{S_O}}\right)\Delta\left(\frac{\mathrm{K[S]}}{1 + \mathrm{K[S]}}\right) \tag{3-5}$$

This still contains a dependence upon substrate concentration but can be simplified further under certain conditions. If K[S] is substantially larger than unity, eq. (3-5) reduces to:

$$\mathrm{LIS} \simeq \left(\frac{\mathrm{L_O}}{\mathrm{S_O}}\right)\Delta \tag{3-6}$$

At low lanthanide–substrate ratios, a significant fraction of the substrate must remain free, so the real requirement is that the association constant, K, be large. It cannot be too large, however, because strong association is inconsistent with a zero value for K_2. Consequently, it is not at all apparent from inspection of these equations when eq. (3-6) provides a satisfactory description of the LIS.

As discussed earlier, the assumption of only 1:1 complex formation may be valid in the case of dpm shift reagents, but fod derivatives normally yield 1:2 adducts as well. If both complexes are taken into account the observed LIS can be described rigorously:

$$\mathrm{LIS} = 2\left(\frac{L_O}{S_O}\right)\Delta_2 + \frac{[LS]}{S_O}\left\{\Delta_1 - 2\Delta_2\left(\frac{1 + K_1[S]}{K_1[S]}\right)\right\} \tag{3-7}$$

Once again, substantial simplification may result at low LSR–substrate ratios. If $[LS]/[S_O]$ is sufficiently small, then the second term of eq. (3-7) becomes negligible in comparison to the first term. The overall equation reduces to:

$$\mathrm{LIS} \simeq 2\left(\frac{L_O}{S_O}\right)\Delta_2 \tag{3-8}$$

The slope of a plot of observed LIS vs. the lanthanide–substrate ratio is equal to twice the bound shift of the 1:2 complex in this situation.[31,61] Despite the apparent simplicity of eq. (3-8), there may be few situations in which it is applicable. The requirement for simplification of eq. (3-7) is that $[LS]/[S_O]$ be very small, and this normally is true only when the lanthanide–substrate ratio is also small. However, if $[L_O]/[S_O]$ is very small, then neither term in eq. (3-7) can be neglected. Inspection of eqs. (3-5) and (3-7) shows that there is actually a dependence of the observed LIS on the concentration of free substrate. Because the experiment is carried out using excess substrate, there also must be a dependence on the total substrate concentration. Any errors in bound shifts determined by this method will result from neglecting the term in eq. (3-5) or (3-7) that contains the dependence on substrate concentration. Consequently, for situations in which the gradient method does not reproduce the true bound shift accurately, the result also depends on the substrate concentration that has been employed in the experiment.

The validity of the gradient method was assessed by evaluating a series of hypothetical compounds (A–D) with different association constants.[61] For each of these compounds, A–D, bound shifts of 76.00 and 46.06 were selected as the bound shifts for the 1:1 and 1:2 complexes, respectively. [These values correspond to the bound shifts determined experimentally with $Yb(dpm)_3$ for the protons adjacent to the carbonyl group of 2-adamantanone].

Each of the hypothetical experiments with the gradient method employed a substrate concentration of 0.15 M, and this was held constant for a series of 13 shift reagent concentrations ranging from 0.0 to 0.06 M. For each of the 13 hypothetical solutions the concentrations of all species in solution were calculated rigorously according to eqs. (3-3) and (3-4). Hypothetical observed LISs were next predicted with eq. (3-2). The final stage of the hypothetical experiment was the analysis of these "observed"

LISs, and they were analyzed in two ways. In both cases the "observed" LISs were plotted as a function of the lanthanide–substrate ratio, but in the second analysis a random error was superimposed on the shifts that had been rigorously calculated according to eqs. (3-3) and (3-4). The random errors were restricted to a maximum of 0.3 ppm, a magnitude that was consistent with our experimental work using 60-MHz NMR spectrometers[61] with $Yb(dpm)_3$ as the shift reagent. The substantial line-broadening effects of Yb frequently make it difficult to assign peak positions more accurately, particularly when there is some overlap of signals from different nuclei. With a maximum LIS of 76 ppm, the typical experimental error in observed chemical shift is quite small (less than 1%) when expressed as a relative error. The absolute errors in chemical shift for europium reagents are usually smaller because there is less line broadening. However, the magnitudes of the induced shifts are smaller as well, so the relative errors in the LIS are comparable. Relative errors on the order of 0.5% have been typical of our extensive studies with $Eu(fod)_3$.[38,42–44,65,66] For these reasons the random errors employed for this analysis should be a good approximation of experimental error in most situations.

The hypothetical experiments were designed to answer two questions. First, is the gradient method capable of extracting the bound shift correctly (for either the 1:1 or 1:2 complex) from the "observed" shifts at different lanthanide concentrations? Second, if it fails as a general procedure, are there limits (where either the 1:1 or 1:2 complex predominates) for which the calculated bound shift corresponds to just one of the complexes?

The results of these hypothetical experiments[61] with compounds A–D using the gradient method are summarized in Table 3-1, and the data demonstrate that the accuracy of the method exhibits a marked dependence on the nature of the substrate. A plot of the data for compound A is illustrated in Figure 3-3. For compound A, which binds strongly to the shift reagent, both K_1 and K_2 are reasonably large. The respective values of 1000 and 200 are comparable to the association constants observed for many carbonyl compounds with $Eu(fod)_3$.[42–44] As discussed above, the experimental conditions of the gradient method favor formation of the 1:2 complex, and eq. (3-8) suggests that the slope of the plot of "observed" LIS vs. $[L_O]/[S_O]$ should be equal to twice the bound shift for the 1:2 complex (ie, 92.2 ppm). In fact, the calculated "bound shift" of 90.2 ppm is quite similar to that value and is clearly very different than the bound shift of 76 ppm for the 1:1 adduct. Nevertheless, the difference of 2 ppm corresponds to an error of more than

TABLE 3-1. BOUND SHIFTS AND ASSOCIATION CONSTANTS USING THE GRADIENT METHOD[a]

Compound	Parameter	Actual values	Calculated values[b]	
			No error	Random error[c]
A	K_1	1000 ($K_2 = 200$)	—	—
	Δ_1	76.00 ($\Delta_2 = 46.06$)	90.18	90.11 ± 0.26
	Bound fractions[d]	LS 0.0–0.05; LS_2 0.0–0.70		
B	K_1	50 ($K_2 = 10$)	—	—
	Δ_1	76.00 ($\Delta_2 = 46.06$)	73.90	74.21 ± 0.41
	Bound fractions[d]	LS 0.0–0.20; LS_2 0.0–0.30		
C	K_1	5 ($K_2 = 1$)	—	—
	Δ_1	76.00 ($\Delta_2 = 46.06$)	31.22	31.15 ± 0.26
	Bound fractions[d]	LS 0.0–0.14; LS_2 0.0–0.04		
D	K_1	5 ($K_2 = 0$)	—	—
	Δ_1	76.00	29.58	29.51 ± 0.26
	Bound fractions[d]	LS 0.0–0.15		

[a] Bound shifts are in ppm; dissociation constants are in M^{-1}.
[b] Calculations are based on the assumption that only the 1:1 complex is formed.
[c] The "observed" LIS had a random error $\leqslant 0.3$ ppm; ranges in the derived bound shifts are $\pm$ standard deviations for three independent determinations.
[d] The fractions of total substrate that are complexed under the conditions of the experiment.

2%, which may be unacceptable even if one is willing to use data corresponding to the 1:2 complex.

For compound B (with $K_1 = 50$ and $K_2 = 10$) the calculated bound shift is 73.9 ppm, a value that is considerably smaller than twice the bound shift for the LS_2 complex, although it differs from that of the 1:1 adduct by only about 3%. The variation in results for compounds A and B reflects the dominance of the LS_2 complex for the former. For the concentration range of shift reagent used in the gradient method, it is predominantly the 1:2 complex that is formed with compound A. The fraction of substrate present as the 1:2 adduct ranges up to 70%, whereas no more than 5% ever exists as the 1:1 adduct. In contrast, both complexes are formed to similar extents with compound B, and this is reflected by a calculated bound shift that is intermediate between the limits of eqs. (3-6) and (3-8). The data in Table 3-1 for compounds C and D further show that the gradient method completely fails to calculate the correct bound shift when there is only weak binding between substrate and shift reagent. Even compound D, which forms the 1:1

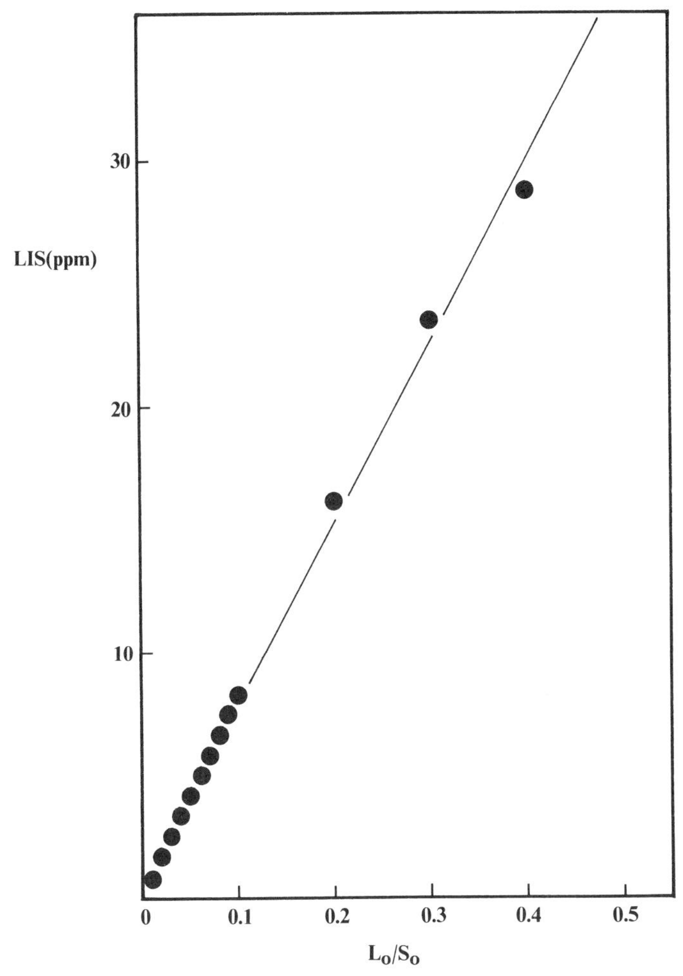

Figure 3-3. Plot of LIS vs. lanthanide–substrate ratio for compound A with the gradient method.

adduct exclusively, yields a measured bound shift that is less than half the actual value.

Why does the gradient method fail to calculate the correct bound shifts? Several factors are involved, but the most important must certainly be the validity of eqs. (3-6) and (3-8). The data for compound A indicate that eq. (3-8) correctly describes the behavior of a strongly binding substrate (at least when the binding constants are comparable to those for compound A). However, substrates that bind more weakly exhibit LISs that are not described by either eq. (3-8) or eq. (3-6). Because this describes a great many situations—especially with dpm reagents—the data in Table 3-1 indicate that the gradient method should be avoided for careful structure analysis with shift reagents.

Reciprocal Method

Another procedure for extracting bound shifts from observed LISs was introduced by Armitage[67,68] in 1971, and Kelsey reported a comparable technique the following year.[69] The method specifically assumes the formation of only a single complex between shift reagent and substrate. Algebraic manipulation of eqs. (3-3) and (3-2) (where the $[LS_2]$ terms have been deleted) affords the following relationship:[61]

$$S_O = (L_O + [LS]^2/S_O)\frac{\Delta}{LIS} - (L_O + 1/K) \qquad (3\text{-}9)$$

At very low $[L_O]/[S_O]$ ratios, the term $[LS]^2/S_O$ must be small in comparison to $[L_O]$, and this results in simplification of eq. (3-9) to the following relationship:

$$S_O \simeq L_O\Delta(1/LIS) - (L_O + 1/K) \qquad (3\text{-}10)$$

The form of eq. (3-10) indicates that if the concentration of substrate (at constant shift reagent concentration) is plotted vs. the reciprocal of the observed LIS, the experimental points should fall on a straight line. The slope of the resulting line then affords the bound shift (slope/$[L_O]$), and the intercept yields the corresponding association constant, ie, $-1/(\text{intercept} + [L_O]$. In contrast to the situation described above for the gradient method, the only approximation used in deriving eq. (3-10) is mathematically sound for the situation in which only a single complex is formed. As long as the substrate binds weakly to the shift reagent (thereby favoring formation of only the 1:1 complex), reasonably accurate bound shifts can be expected from the reciprocal method. There does not appear to be any simple relationship analogous to eq. (3-10), however, that describes the behavior of the many systems that yield both 1:1 and 1:2 adducts.[61] Consequently, two questions must be answered with regard to the reciprocal method. First, will the correct bound shifts and association constants be obtained when experiments are carried out with a substrate that forms only a 1:1 complex? Second, what will be the results when the method is applied to systems yielding two complexes?

The hypothetical compounds A–D again provide a convenient way to assess the accuracy and reliability of the experimental technique, and

Table 3-2 summarizes the results for a series of computer-simulated experiments with the reciprocal method. In accord with the procedure suggested by Armitage[67,68] the shift reagent concentration was held constant at 0.005 M, and the substrate concentration was varied from 0.03 to 0.20 M (for a total of seven data points). As described for the gradient method, "observed" LISs were calculated rigorously with eqs. (3-2)–(3-4) for each of the compounds. The data in Table 3-2 are derived from analysis of the linear regression. Analysis of substrate concentration vs. "observed" LIS according to eq. (3-10) then yielded the data summarized in Table 3-2. The data for compound A also are illustrated in the form of the appropriate plot in Figure 3-4.

The bound shifts obtained with the reciprocal method fall in a narrower range than do the values determined with the gradient method. Nevertheless, there are some similarities. The bound shift calculated for compound A again seems to correspond to a value that is twice that for the 1:2 complex, and this is verified by inspection of the bound fractions (footnote *d* of the table). Under the conditions of the experiment, compound A yields the 1:2 complex almost exclusively, so the calculated

TABLE 3-2. Bound Shifts and Association Constants Using the Reciprocal Method[a]

			Calculated values[b]	
Compound	Parameter	Actual values	No error	Random error[c]
A	K_1	1000 ($K_2 = 200$)	383	38 ± 210
	Δ_1	76.00 ($\Delta_2 = 46.06$)	93.61	94.76 ± 4.79
	Bound fractions[d]	LS < 0.01–0.03; LS_2 0.05–0.27		
B	K_1	50 ($K_2 = 10$)	47.0	57.9 ± 15.7
	Δ_1	76.00 ($\Delta_2 = 46.06$)	95.07	96.66 ± 6.33
	Bound fractions[d]	LS 0.01–0.08; LS_2 0.03–0.05		
C	K_1	5 ($K_2 = 1$)	4.5	20.8 ± 11.1
	Δ_1	76.00 ($\Delta_2 = 46.06$)	89.14	63.96 ± 25.73
	Bound fractions[d]	LS 0.01–0.02; LS_2 < 0.01		
D	K_1	5 ($K_2 = 0$)	5.3	37.7 ± 24.1
	Δ_1	76.00	72.95	51.49 ± 19.87
	Bound fractions[d]	LS 0.01–0.02		

[a] Bound shifts are in ppm; dissociation constants are in M^{-1}.
[b] Calculations are based on the assumption that only the 1:1 complex is formed.
[c] The "observed" LIS had a random error ⩽0.3 ppm; ranges in the derived association constants and bound shifts are ± standard deviations for three independent determinations.
[d] The fractions of total substrate that are complexed under the conditions of the experiment.

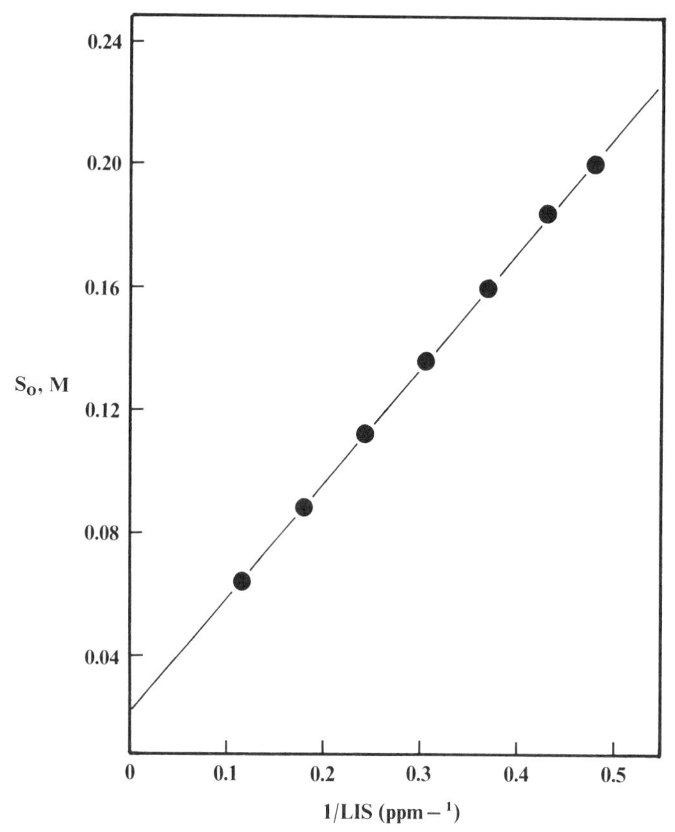

Figure 3-4. Plot of 1/LIS vs. substrate concentration (at constant shift reagent concentration) with the reciprocal method for compound A.

bound shift must reflect that adduct. Clearly eq. (3-10) and the reciprocal method are not applicable for such substrates as compound A that bind strongly with shift reagent. For compounds B, C, and D the bound shifts calculated by the reciprocal method improve steadily. Only in the case of D (which represents a limit, rather than a real situation), however, does the calculated bound shift agree well with the actual value. Apparently any 1:2 complex formation at all perturbs the "observed" LIS to a sufficient extent that an inaccurate bound shift results.

The association constants calculated by the reciprocal method in the absence of experimental error are in fairly good agreement with the actual values, although it is the larger association constant (ie, K_1) that is obtained in each case. The agreement is therefore only coincidental, and one must be careful in interpreting these results.

The superimposition of a random error on the "observed" LIS has a dramatic effect on the bound shifts and association constants calculated by the reciprocal method. In large part this reflects the small induced shifts that are produced in this procedure. If the substrate binds weakly (the only situation for which the method can be expected to afford good results), then only a small fraction is present in complexed form. The bound fractions reported in Table 3-2 show that under 10% of compound B can be complexed during most of the experiment, and a total of about 2% complexation is the upper limit for compound C. This in turn means that the maximum induced shifts can be only about 8 and 2 ppm, respectively (even though the bound shift is 76 ppm for the 1:1 adduct). The simulated experimental errors result in very large errors in the bound shifts calculated by the reciprocal method for compounds C and D. Because random experimental errors do not affect different nuclei in a substrate to the same extent (or even in the same direction), the results in Table 3-2 argue against the use of the reciprocal method for rigorous shift reagent work. The association constants determined by this method are fairly accurate for compound B, but random experimental error again seems to cause large uncertainties for the other compounds. This is not surprising if one considers that these association constants are calculated from the reciprocal of a rather small number. Errors in the calculated intercept for eq. (3-10) therefore are magnified in the association constants.

Equimolar Method

For the situation in which only a 1:1 complex is formed, a remarkable simplification of eqs. (3-2) and (3-3) results when $[S_O]$ is equal to $[L_O]$. Algebraic manipulation[60,61,69] of eqs. (3-2) and (3-3) (with the $[LS_2]$ terms deleted) affords the following mathematical description of the LIS:

$$\mathrm{LIS} = \Delta_1 - \left(\frac{\mathrm{LIS}}{S_O}\right)^{1/2}\left(\frac{\Delta_1}{K_1}\right)^{1/2} \tag{3-11}$$

An experimental procedure for determining bound shifts using equimolar quantities of shift reagent and substrate was developed by Bouquant and Chuche,[70] and this also was evaluated by simulated experiments with compounds A–D. A plot of the observed LIS vs. the square root of LIS/S_O is linear (assuming only the 1:1 complex is formed). The line has an intercept equal to the bound shift, and the slope is equal to the square root of the bound shift divided by the association constant.

The equimolar method was also analyzed using computer-simulated experiments with compounds A–D,[61] and Figure 3-5 illustrates the plot

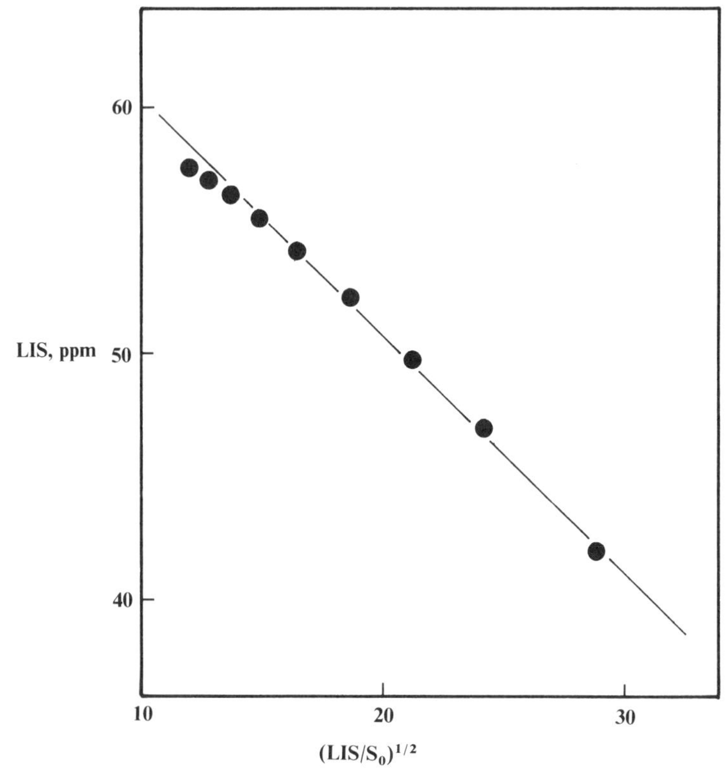

Figure 3-5. Plot of LIS vs. $(LIS/S_O)^{1/2}$ with the equimolar method for compound A.

obtained with compound A. All of the simulated experiments employed equal concentrations of LSR and substrate, and nine data points were used over a concentration range from 0.05 to 0.40 M. The results of these simulated experiments are summarized in Table 3-3.

Inspection of Table 3-3 reveals that the agreement between "experimental" and actual bound shifts is poor for compound A but improves steadily for compounds B, C, and D. For the latter two compounds excellent agreement is also seen between the "experimental" and the actual association constants. In all cases the effect of random "experimental" error on the calculated values is only minor. The data in Table 3-3 clearly show that the equimolar method provides accurate bound shifts and equilibrium constants as long as the substrate does not bind strongly to the shift reagent. It is not really advisable to place a specific upper limit on the magnitude of K_1 that will yield accurate results by the equimolar method. However, the discrepancy between actual and "experimental" bound shifts has probably become too large for many

TABLE 3-3. Bound Shifts and Association Constants Using the Equimolar Method[a]

Compound	Parameter	Actual values	Calculated values[b]	
			No error	Random error[c]
A	K_1	1000 ($K_2 = 200$)	10,112	9886 ± 215
	Δ_1	76.00 ($\Delta_2 = 46.06$)	62.63	62.67 ± 0.11
	Bound fractions[d]	LS 0.53–0.53; LS_2 0.43–0.47		
B	K_1	50 ($K_2 = 10$)	82.1	83.8 ± 2.0
	Δ_1	76.00 ($\Delta_2 = 46.06$)	68.89	68.73 ± 0.30
	Bound fractions[d]	LS 0.45–0.52; LS_2 0.17–0.38		
C	K_1	5 ($K_2 = 1$)	5.0	5.0 ± 0.1
	Δ_1	76.00 ($\Delta_2 = 46.06$)	78.45	78.74 ± 0.58
	Bound fractions[d]	LS 0.17–0.43; LS_2 0.14–0.15		
D	K_1	5 ($K_2 = 0$)	5.0	5.0 ± 0.1
	Δ_1	76.00	76.00	75.74 ± 0.45
	Bound fractions[d]	LS 0.17–0.50		

[a] Bound shifts are in ppm; dissociation constants are in M^{-1}.
[b] Calculations are based on the assumption that only the 1:1 complex is formed.
[c] The "observed" LIS had a random error $\leqslant 0.3$ ppm; ranges in the derived association constants and bound shifts are ± standard deviations for three independent determinations.
[d] The fractions of total substrate that are complexed under the conditions of the experiment.

purposes with compound B, where the actual and calculated bound shifts differ by approximately 10%. Although most common oxygen- and nitrogen-containing functional groups exhibit association constants that are on the order of 50 M^{-1} or larger with $(fod)_3$ reagents in CCl_4, there are two ways in which smaller association constants can be obtained. The first involves changing to a more polar solvent, such as deuteriochloroform,[65] and this can lead to a reduction of the K_1 by an order of magnitude. Nevertheless, the available data for $Eu(fod)_3$ studies[42–44] suggest that many oxygen- and nitrogen-containing compounds still exhibit association constants that are too large for reliable use with the equimolar method. The second way in which smaller association constants can be obtained is through the use of dpm reagents. Rackham's extensive studies[71–73] indicate that most carbonyl compounds (ie, esters, aldehydes, ketones) have association constants for interaction with dpm reagents in $CDCl_3$ that are on the order of 10 M^{-1} or less. On the other hand, many amines and alcohols have association constants that are too large, as do many sulfoxides, amine oxides, and phosphine oxides. In addition, our previous studies[66,74–78] with nitriles using $Eu(fod)_3$ indicate that these also bind rather weakly with dpm reagents and consequently are likely to afford good results using the equimolar method.

The data in Table 3-3 bring up an additional question. Why is the calculated bound shift for compound A so close to that of the 1:1 complex, when the 1:1 and 1:2 adducts are formed to nearly the same extents? Similarly, the 1:2 adduct provides a substantial contribution to the total fraction of complexed substrate for compounds B and C, yet the bound shifts calculated using eq. (3-11) correspond closely to that of the 1:1 complex and are very different from the 46 ppm of the 1:2 adduct. The answer seems to relate to the typical ratio of the bound shifts for the two complexes. The bound shift for the 1:2 complex is usually about 60% of that for the 1:1 adduct, and we have shown[61] previously that an approximate relationship equivalent to eq. (3-11) can be described for situations in which both complexes are formed.

Two-Step Method

The two-step method was first introduced by Shapiro and Johnston,[11] and analogous procedures were developed subsequently by Reuben[14] and Inagaki.[79,80] This technique employs a rigorous mathematical analysis of the LIS according to eqs. (3-1)–(3-4), and no assumptions are made (other than to accept the model described by Figure 3-1). Moreover, in contrast to other methods, there are no approximations regarding measurement of total concentrations of shift reagent and substrate. The sample (containing both shift reagent and substrate) is diluted with a stock solution of substrate, and concentrations are determined accurately by weighing the sample after each dilution. The importance of accurate measurement of volumes is partly a result of the high molecular weight of lanthanide shift reagents. A 0.5-ml sample of a solution that is 0.2 M in both shift reagent and substrate, contains ca. 0.01 g of a typical substrate but 0.1 g of shift reagent. Clearly the contribution of shift reagent to the total volume cannot be neglected.

Four variables (the bound shifts for the 1:1 and 1:2 complexes together with the association constants K_1 and K_2) are optimized by regression analysis. The "best" values of these four parameters are those leading to the smallest discrepancies between experimental LISs and the LISs calculated with eqs. (3-2)–(3-4). Because the method is rigorous, the true bound shifts and association constants are expected to be obtained without difficulty. However, Deranleau[81] and Reuben[82] have shown that the accuracy with which such equilibrium parameters can be obtained depends highly on the extent of complexation in the samples used to obtain experimental data. The fitting procedure is most reliable when data points are obtained over the range of 20–80% complexation for each complex that is formed. This requires that a larger number of data points be obtained—from solutions with excess shift reagent (favoring 1:1 adduct) to ones with excess substrate (favoring 1:2 adduct). When spectra are obtained for as many as 25 different concentration ratios,[11,31] the two-step method is experimentally tedious. It is therefore important to determine whether this inconvenience is compensated by an improvement in accuracy (relative to the gradient, reciprocal, and equimolar methods, or in comparison to a modified[83] two-step procedure). A second question to be answered is whether the two-step method provides accurate results when K_2 is very small. Is any simplification introduced by setting K_2 equal to zero as an approximation in such cases?

Simulated experiments were again carried out for compounds A–D.[61] The procedure of Shapiro and Johnston[11,31] was followed, and LISs were rigorously predicted for 25 different concentrations of shift reagent. Substrate concentration was held constant at 0.2 M, and shift reagent concentration was varied from 0.0 to 0.6 M. Figure 3-6 shows a typical plot of the data obtained in this way, and the results of the simulated experiments are summarized in Table 3-4.

For compounds A and B the two-step procedure affords more accurate bound shifts than any of the other three methods (and of course it is the only procedure that yields bound shifts for both the 1:1 and 1:2 complexes). The association constants show typical deviations of 10–30% from the true values, and this is consistent with the errors reported previously in our experimental studies.[42–44,65,66] The association constant calculated for compound B using the reciprocal method was closer to the true value than that calculated by the two-step procedure, but this was probably a coincidence. In general, the reciprocal method affords unreliable LIS parameters, particularly when experimental error is taken into account. Clearly, the two-step method is the only available procedure that affords reliable bound shifts and association constants when there is a strong interaction between substrate and shift reagent. A major reason for the superiority of the two-step method can be seen by inspecting the bound fractions shown in Table 3-4. In distinct contrast to the results with the other methods, a large range of complexation (both 1:1 and 1:2 adducts) is found for compound A. For compound B, a smaller range is covered for the 1:2 complex, but the adduct of primary interest (the 1:1 complex) is still well determined.[81,82] Although it is again difficult to set specific limits, the two-step method is probably

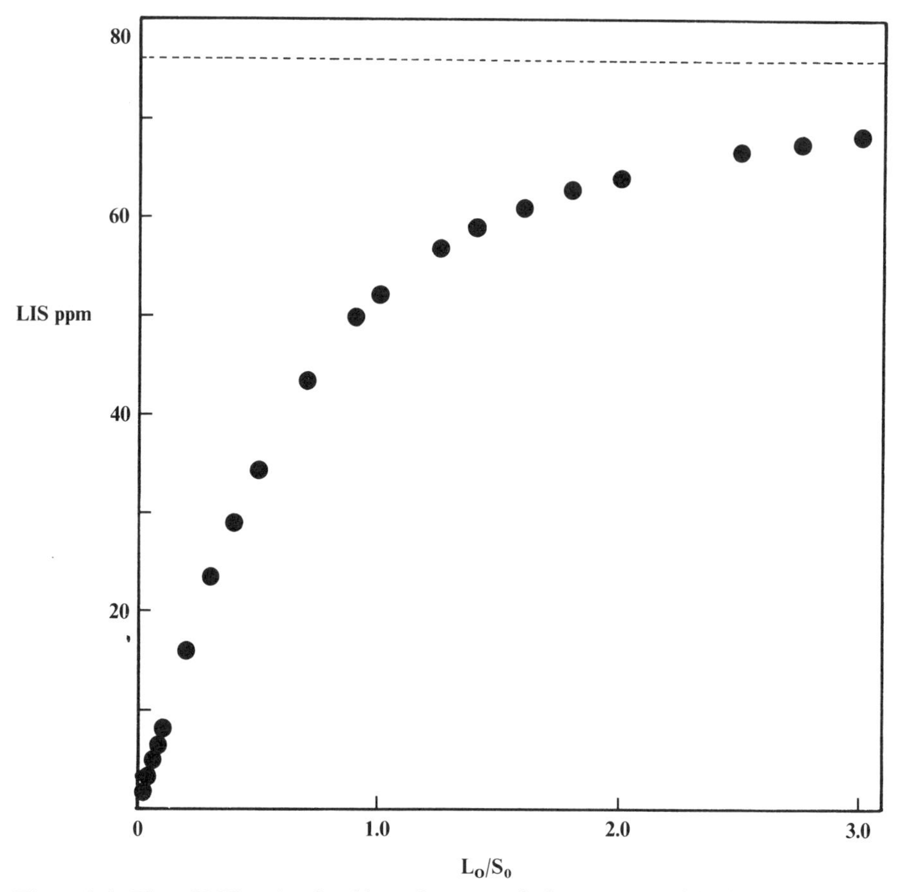

Figure 3-6. Plot of LIS vs. lanthanide–substrate ratio for compound A with the two-step method. The dotted line indicates the actual value of Δ_1 (which is approached asymptotically by the observed LIS).

well suited to most studies of dpm reagents with strong donors, such as amines and sulfoxides, and to nearly all uses of fod reagents. In most such cases K_1 has been found to have a value of 50 M^{-1} or larger.[42–44,71–73]

The results for the two-step method with compounds C and D are less clear-cut. The bound shift and association constant for compound D are quite accurately determined, but it must be recalled that this case represents a limit rather than a real situation. For a substrate that binds weakly to shift reagent compound C is a much better model, and here the two-step method gives only marginally satisfactory results. When the normal procedure is employed so that the "best" values of all four variable parameters are sought, the values for the 1:2 complex are totally

TABLE 3-4. Bound Shifts and Association Constants Using the Two-Step Method[a]

Compound	Parameter	Actual values	Calculated values: No error	Calculated values: Random error[b]
A	K_1	1000	1057	979 ± 95
	K_2	200	211	203 ± 19
	Δ_1	76.00	76.96	76.21 ± 0.11
	Δ_2	46.06	46.04	46.08 ± 0.06
	Bound fractions[c]	LS 0.0–0.86; LS_2 0.0–0.77		
B	K_1	50	36.7	38.5 ± 3.0
	K_2	10	5.8	7.7 ± 1.4
	Δ_1	76.00	75.49	76.02 ± 0.30
	Δ_2	46.06	58.37	54.21 ± 4.39
	Bound fractions[c]	LS 0.0–0.82; LS_2 0.0–0.33		
C	K_1	5	6.7[d,e]	6.2 ± 1.1[d,f]
	K_2	1	—	—
	Δ_1	76.00	69.13[d,g]	69.14 ± 1.18[d,h]
	Δ_2	46.06	—	—
	Bound fractions[c]	LS 0.0–0.60; LS_2 0.0–0.07		
D	K_1	5	5.0	5.0 ± 0.1
	K_2	0[i]		
	Δ_1	76.00	76.06	76.13 ± 0.65
	Bound fractions[c]	LS 0.0–0.64		

[a] Bound shifts are in ppm; dissociation constants are in M^{-1}.
[b] The "observed" LIS had a random error ⩽ 0.3 ppm; ranges in the derived association constants and bound shifts are ± standard deviations for three independent determinations.
[c] The fractions of total substrate that are complexed under the conditions of the experiment.
[d] K_2 is set equal to zero.
[e] If K_2 is optimized, values of 1.2 and 3.2 are obtained for K_1 and K_2, respectively.
[f] If K_2 is optimized, values of 1.0 ± 0.1 and 4.0 ± 0.7 are obtained for K_1 and K_2, respectively.
[g] If K_2 is optimized, values of 62.44 and 203.02 are obtained for Δ_1 and Δ_2, respectively.
[h] If K_2 is optimized, values of 52.12 ± 0.77 and 221.04 ± 19.07 are obtained for Δ_1 and Δ_2, respectively.
[i] No 1:2 complex is formed for compound D.

unreasonable. Both the bound shift for the 1:2 adduct and the association constant for its formation are computed to be larger than the corresponding values for the 1:1 complex. For this reason K_2 was set at zero, and the analysis was restricted to a one-step model (ie, to formation of only the 1:1 adduct). This approximation afforded a bound shift and association constant that was slightly improved, but there was still a discrepancy of about 10% in the bound shift. The less satisfactory results with compound C appear to arise from the small range of 1:2

adduct formation. This complex is formed only to a maximum of 7%, which is not sufficient to determine the corresponding LIS parameters properly in the regression analysis. Yet the approximate method (in which K_2 is set equal to zero) introduces a new error, because all of the induced shift is incorrectly ascribed to the 1:1 complex (ignoring the contribution of up to 7% from the 1:2 adduct). This apparently causes the fitting procedure to select as "best" value an association constant that is larger than the true K_1 and a bound shift that is smaller than the true value.

For compound C it seems clear that the equimolar method provides results superior to those of the two-step procedure. Why is this so when the analysis of compound C by either method employs the incorrect assumption that only the 1:1 complex is formed? The failure of the two-step method seems to result from two factors. First, as just stated, the optimization procedure is attempting to fit contributions to the "observed" LIS from the 1:2 adduct to a set of equations that consider only the 1:1 complex. Second, the optimization procedure is based on the minimization of the variance (ie, the discrepancy between "observed" and "predicted" LIS), and this is relatively insensitive to variations in the magnitude of the association constants. As a result, small random (experimental) errors in the "observed" LIS can result in a larger variance for the true values of the LIS parameters than for some other values. Stated differently, the fitting procedure is rather "soft," and random errors can result in calculated bound shifts and association constants that differ significantly from the true values. This is only a minor problem when the complexes of interest are well defined experimentally (ie, the full range of complexation is observed). For this reason the two-step method affords good results with strongly complexing systems such as compounds A and B.

More recently, we have investigated the use of the two-step method for a smaller number of data points.[83] Although the results are more sensitive to experimental errors, rather accurate bound shifts and association constants were obtained, even when spectra were measured at only six or seven different concentration ratios.

Conclusions and Recommendations

The previous sections of this chapter have presented a review of shift reagent equilibria, together with a critical analysis of the various methods for experimental determination of bound shifts. This section summarizes the advantages and disadvantages of the various methods and uses that information to make recommendations about selecting the best method for a particular compound. The goal is not to promote a particular methodology, but to determine what procedure is most convenient and most reliable in a given set of circumstances. Several suggestions also are made for modifying these methods that may result in more convenient experimental procedures.

Equimolar and Two-Step Methods

If accurate bound shifts and association constants are desired, then either the two-step or the equimolar method should be used. The two procedures are complementary, and each provides optimum results for only certain situations. When there is strong binding between substrate and shift reagent (ie, K_1 is on the order of 50 M^{-1} or larger), the two-step method is superior. The experimental procedure utilizes lanthanide–substrate ratios that range from excess lanthanide to excess shift reagent, and this allows both the 1:1 and the 1:2 complexes to be well defined. When weaker binding is observed between substrate and shift reagent (ie, K_1 is on the order of 5 M^{-1} or smaller), the equimolar method affords superior values for the bound shift and association constant. Although this procedure employs the approximation that only a single complex is formed, the approximation is a reasonably good one. Contributions from the 1:2 complex decrease steadily as the association constant becomes smaller, and the reciprocal method is most reliable when the association constant is quite small. When K_1 is between 5 and 50 M^{-1}, either method appears to be satisfactory. How can the magnitude of the association constant be known before the experiment is carried out? A reasonable estimate can be made on the basis of data that have been reported in the literature, and Tables 3-5 through 3-7 present a group of selected association constants for this purpose.

We have attempted to select those association constants most useful in predicting the behavior of other compounds, so the tables do not provide an exhaustive compilation. Additional data can be found in many of the references cited in the tables as well as in several additional articles.[99–103] Although only rough estimates of association constants can be made for new compounds, there are several useful trends in the data of Table 3-5. First of all, there is a direct correlation between the binding constant for a particular substrate and its basicity (as reflected by its proton affinity).[42] There are some exceptions to this—for example, sulfur derivatives exhibit proton affinities that are stronger than those

TABLE 3-5. ASSOCIATION CONSTANTS FOR REPRESENTATIVE ORGANIC COMPOUNDS WITH LANTHANIDE SHIFT REAGENTS

	Shift reagent	Solvent	K (M^{-1})	Ref.
Ketones and aldehydes				
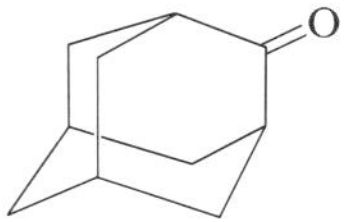 2-Adamantanone (structure)	$Yb(dpm)_3$	CCl_4	5	61
	$Eu(fod)_3$	CCl_4	2,300 ± 1,400	43, 65
	$Eu(dpm)_3$	$CDCl_3$	5	73f
	$Eu(fod)_3$	$CDCl_3$	316	65
1-Adamantyl–$\overset{O}{\overset{\|\|}{C}}CH_3$ (structure)	$Eu(dpm)_3$	$CDCl_3$	14	73b
	$Eu(fod)_3$	CCl_4	5,400	42
	$Eu(fod)_3$	CCl_4	56	42
$CH_3—\overset{O}{\overset{\|\|}{C}}—CH_3$	$Eu(fod)_3$	CCl_4	1,000	43
	$Eu(dpm)_3$	C_6H_{12}	218	84
$CH_3—\overset{O}{\overset{\|\|}{C}}—CH_2—CH_3$	$Eu(fod)_3$	CCl_4	250	85
	$Eu(dpm)_3$	C_6H_{12}	205	84
$C_6H_5—\overset{O}{\overset{\|\|}{C}}—CH_3$	$Eu(dpm)_3$	$CDCl_3$	10	73a
	$Eu(dpm)_3$	C_6H_{12}	210	84
$C_6H_5—\overset{O}{\overset{\|\|}{C}}—H$	$Eu(dpm)_3$	$CDCl_3$	10	73a

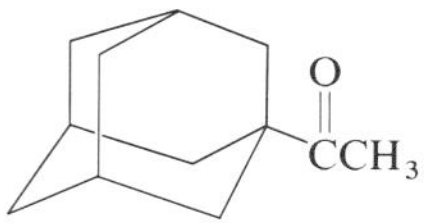

(continued)

TABLE 3-5. *(continued)*

	Shift reagent	Solvent	K (M^{-1})	Ref.
O	$Eu(dpm)_3$	$CDCl_3$	10	73f
	$Eu(fod)_3$	CCl_4	6,600	43
O	$Eu(dpm)_3$	80% CCl_4–20% C_6D_6	170	70b
	$Eu(fod)_3$	CCl_4	4,900	43
	$Ho(dpm)_3$	CCl_4	56	86
Carboxylic acid derivatives (esters, amides)				
$CH_3O—\overset{O}{\overset{\Vert}{C}}—N(CH_3)_2$	$Eu(fod)_3$	CCl_4	1,600	87
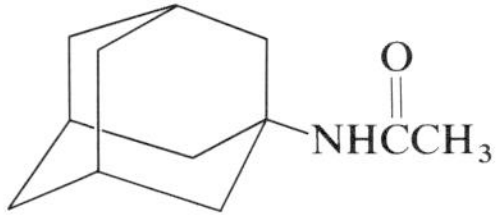 $NH\overset{O}{\overset{\Vert}{C}}CH_3$	$Eu(fod)_3$	CCl_4	1,300	42
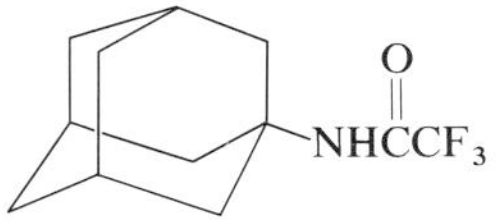 $NH\overset{O}{\overset{\Vert}{C}}CF_3$	$Eu(fod)_3$	CCl_4	140	42
$CH_3—\overset{O}{\overset{\Vert}{C}}—N(CH_3)_2$	$Pr(fod)_3$	$ClCH_2CH_2Cl$	6×10^7	57
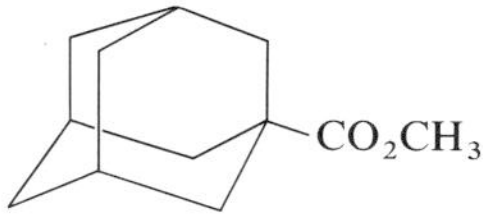 CO_2CH_3	$Eu(dpm)_3$	$CDCl_3$	18	73b

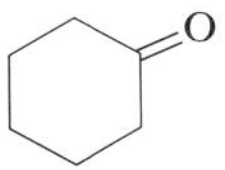

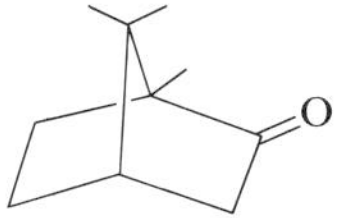

(continued)

TABLE 3-5. *(continued)*

	Shift reagent	Solvent	K (M^{-1})	Ref.
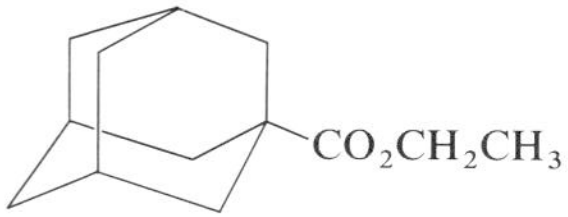 $CO_2CH_2CH_3$	$Eu(fod)_3$	CCl_4	3,600	42
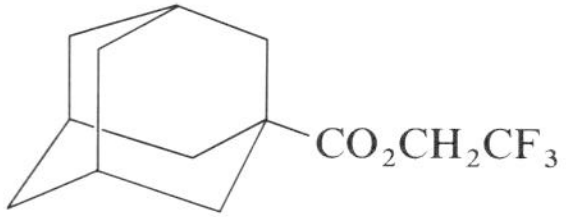 $CO_2CH_2CF_3$	$Eu(fod)_3$	CCl_4	74	42
$C_6H_5CO_2CH_3$	$Eu(dpm)_3$	$CDCl_3$	9	73b
$CH_3CH_2CH_2OC(=O)CF_3$	$Eu(fod)_3$	CCl_4	0.3	88
Alcohols				
$CH_3-CH(OH)-CH_2-CH_3$	$Eu(dpm)_3$	CCl_4	540 (35°C)	89
			870 (22°C)	90
$CH_3-C(CH_3)_2-OH$	$Eu(dpm)_3$	CCl_4	387 (35°C)	89
			760 (22°C)	90
	$Eu(dpm)_3$	$CDCl_3$	200	70b
$C_6H_5CH_2OH$	$Eu(dpm)_3$	$CDCl_3$	76	72
	$Yb(dpm)_3$	$CDCl_3$	15	72
$C_6H_5CH(OH)-C(CH_3)_3$	$Eu(fod)_3$	$CDCl_3$	16	62

(continued)

TABLE 3-5. *(continued)*

	Shift reagent	Solvent	K (M^{-1})	Ref.
(fluorenol structure: H, OH)	$Eu(dpm)_3$	$CDCl_3$	93	72
	$Yb(dpm)_3$	$CDCl_3$	9	72
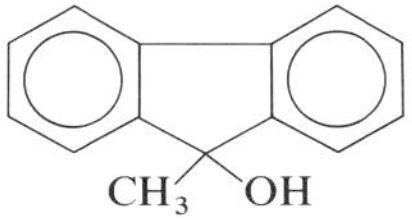 (9-methylfluorenol structure: CH_3, OH)	$Eu(fod)_3$	$CDCl_3$	105	91
(adamantyl structure: CH_2OH)	$Eu(fod)_3$	CCl_4	6,500	42
	$Eu(dpm)_3$	$CDCl_3$	297	73b
Ethers				
$CH_3CH_2CH_2CH_2OCH_3$	$Pr(fod)_3$	CCl_4	30	36
$C_6H_5OCH_3$	$Eu(dpm)_3$	$CDCl_3$	0	73f
	$Eu(fod)_3$-d_{27}	$CDCl_3$	0	73f
	$Yb(fod)_3$-d_{27}	$CDCl_3$	9	73f
	$Pr(fod)_3$	CCl_4	2.5	36
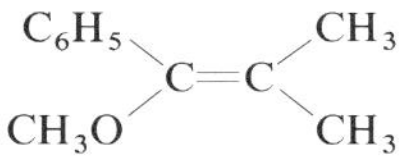 $(C_6H_5)(CH_3O)C{=}C(CH_3)_2$	$Eu(dpm)_3$	CCl_4	3	92
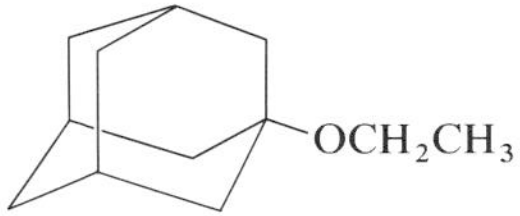 (adamantyl structure: OCH_2CH_3)	$Eu(fod)_3$	CCl_4	6	42

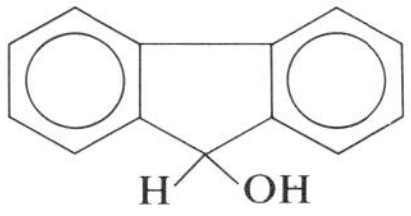

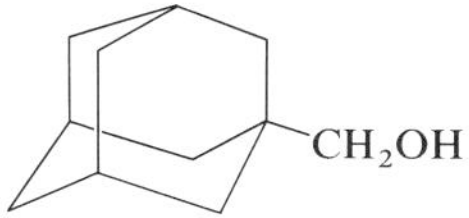

(continued)

TABLE 3-5. *(continued)*

	Shift reagent	Solvent	K (M^{-1})	Ref.
$CH_3CH_2CH_2OSi(CH_3)_3$	$Eu(fod)_3$	$CDCl_3$	1	88
$CH_3CH_2CH_2OSi(CH_3)_2t\text{-}C_4H_9$	$Eu(fod)_3$	$CDCl_3$	0.02	88
$C_6H_5—C(O)—CH_2$ (epoxide)	$Eu(dpm)_3$	$CDCl_3$	13	73f
	$Eu(fod)_3$	$CDCl_3$	45	73f
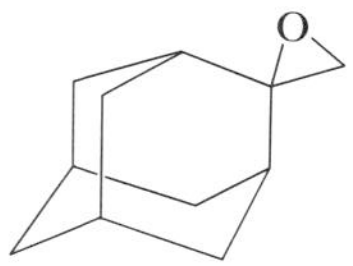 O (spiro-epoxide)	$Eu(fod)_3$	CCl_4	600	93
Sulfoxides, amine oxides, phosphine oxides				
$(C_6H_5)_3P{=}O$	$Ho(dpm)_3$	CCl_4	30,000	86
$(C_6H_5O)_3P{=}O$	$Eu(dpm)_3$	$CDCl_3$	23	73d
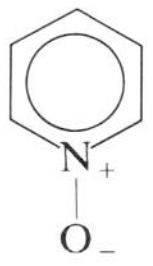 $N^+—O^-$	$Eu(dpm)_3$	$CDCl_3$	1,680	73d
$CH_3—S(=O)—CH_3$	$Eu(dpm)_3$	$CDCl_3$	553	73d
	$Eu(fod)_3$	CCl_4	625	14

(continued)

TABLE 3-5. *(continued)*

	Shift reagent	Solvent	K (M^{-1})	Ref.
Amines				
N	$Eu(dpm)_3$	$CDCl_3$	333	73c
	$Yb(dpm)_3$	C_6H_6	1,380	94
	$Eu(dpm)_3$	C_6H_6	3,890	94
	$Eu(fod)_3$	C_6H_6	10,000	94
N, CH_3	$Eu(dpm)_3$	$CDCl_3$	83	73c
	$Yb(dpm)_3$	C_6H_6	144	94
	$Eu(dpm)_3$	C_6H_6	224	94
	$Eu(fod)_3$	C_6H_6	316	94
N	$Eu(dpm)_3$	$CDCl_3$	9	73e
$C_6H_5CH_2NH_2$	$Eu(dpm)_3$	$CDCl_3$	1,400	73b
NH	$Eu(dpm)_3$	$CDCl_3$	1,360	73c
NH, CH_3	$Eu(dpm)_3$	$CDCl_3$	314	73c
NCH_3	$Eu(dpm)_3$	$CDCl_3$	17	73c
NH_2	$Yb(dpm)_3$	$CDCl_3$	22	71
	$Eu(dpm)_3$	$CDCl_3$	49	73c
	$Eu(fod)_3$	CCl_4	382	71

(continued)

TABLE 3-5. *(continued)*

	Shift reagent	Solvent	K (M^{-1})	Ref.
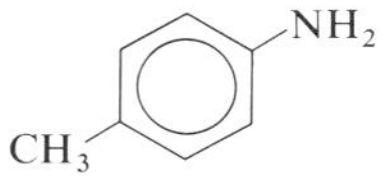 CH_3, NH_2	$Eu(fod)_3$ $Eu(dpm)_3$	$CDCl_3$ CCl_4	212 30	95 71
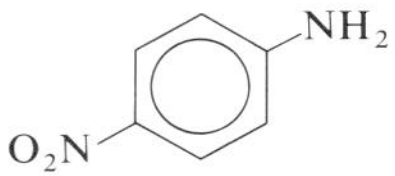 O_2N, NH_2	$Eu(fod)_3$	$CDCl_3$	3	95
$(C_6H_5)_2NH$	$Eu(dpm)_3$	$CDCl_3$	<0.5	73c
Nitriles				
CH_3CN	$Eu(fod)_3$	CCl_4	52	44
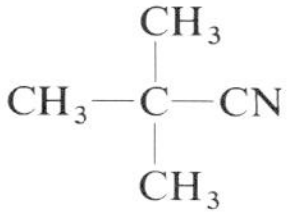 CH_3—C(CH_3)(CH_3)—CN	$Eu(fod)_3$	CCl_4	134	44
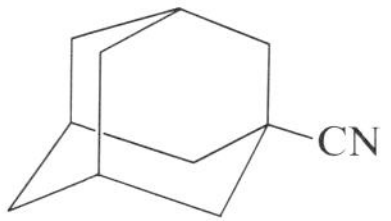 CN	$Eu(fod)_3$	CCl_4	484	44
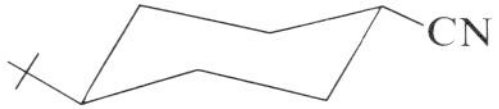 CN	$Eu(fod)_3$	CCl_4	97	44
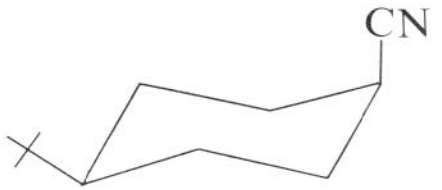 CN	$Eu(fod)_3$	CCl_4	115	44
C_6H_5CN	$Eu(fod)_3$	CCl_4	19	44

(continued)

TABLE 3-5. *(continued)*

	Shift reagent	Solvent	K (M^{-1})	Ref.
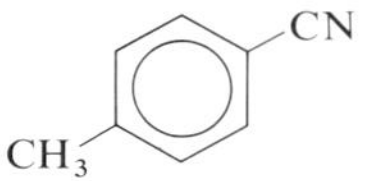 CH_3, CN (benzene ring)	$Eu(fod)_3$	CCl_4	57	44
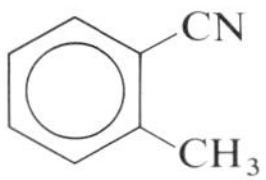 CN, CH_3 (benzene ring)	$Eu(fod)_3$	CCl_4	51	44
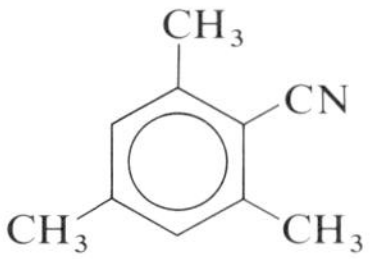 CH_3, CN, CH_3, CH_3 (benzene ring)	$Eu(fod)_3$	CCl_4	68	44
Difunctional (chelating) compounds				
$C_6H_5-CH(OH)-CH(C_6H_5)-C(=O)-OCH_3$	$Eu(fod)_3$	CCl_4	3,600 threo) 4,700 (erythro)	85 85
CH_2OCH_3 $CH_2O(CH_2)_7CH_3$	$Pr(fod)_3$	CCl_4	66,000	36
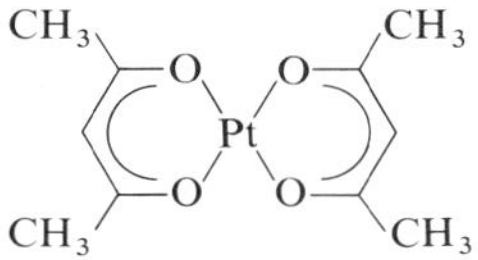 CH_3, CH_3, O, O, Pt, O, O, CH_3, CH_3	$Pr(fod)_3$	$CDCl_3$	10	96

(continued)

TABLE 3-5. *(continued)*

		Shift reagent	Solvent	K (M^{-1})	Ref.
M = Co	R,R′ = CH_3	$Eu(fod)_3$	C_6H_6	10,000	97
		$Eu(fod)_3$	$CDCl_3$	820	52
M = Co	R = t-C_4H_9	$Eu(fod)_3$	$CDCl_3$	830	52
M = Rh	R,R′ = CH_3	$Eu(fod)_3$	$CDCl_3$	26	98
M = PR	R = t-C_4H_9	$Pr(fod)_3$	C_6H_{14}	5,130	34
	R′ = C_3F_7	$Pr(fod)_3$	C_6H_6	10^2–10^3	35
(ie, PrFOD)		$Pr(fod)_3$	CCl_4	140	87
M = Eu	R = t-C_4H_9	$Eu(fod)_3$	CCl_4	367	87
	R′ = C_3F_7	$Eu(fod)_3$	C_6H_6	61	35
(ie, EuFOD)					

TABLE 3-6. Association Constants and Bound Shifts for a Series of Substituted Cyclohexanols

Compound	K_1 (M^{-1}) and (Δ_1, ppm)[a]			
	$Yb(dpm)_3/CDCl_3$[b]	$Yb(dpm)_3/CCl_4$[b]	$Eu(dpm)_2/CDCl_3$[b]	$Eu(fod)_3/CCl_4$[c]
H, OH	24	150	301 (24.8)	—
CH_3, OH	—	25 (57.0)	94 (16.3)	—
OH, H	32	150	324 (26.5)	10,000 (22.6)
H, OH	23	150	243 (24.3)	5,345 (22.6)
OH, CH_3	—	16 (56.5)	70 (16.6)	209 (18.4)
CH_3, OH	—	39 (54.1)	124 (16.0)	614 (15.4)

[a] Bound shifts are for the hydrogens (H or CH_3) on C-1 of the cyclohexane ring.
[b] Ref. 114
[c] Ref. 31

TABLE 3-7. Solvent Effects on Association Constants

Compound	Shift Reagent	K_1 (M^{-1})							
		C_7H_{14}[a]	C_6H_{12}[b]	CCl_4	C_6H_6	CS_2	$CDCl_3$	CH_2Cl_2	Ref.
$Eu(dpm)_3$	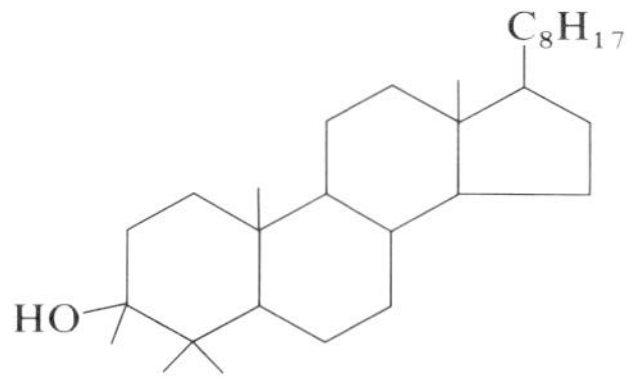 HO, C_8H_{17}	—	854	330	188	303	62	—	114,115
$Eu(dpm)_3$	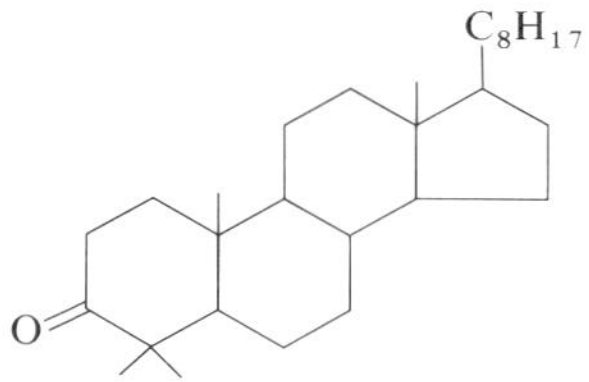 O, C_8H_{17}	—	339	122	70	101	27	—	115
$Eu(fod)_3$	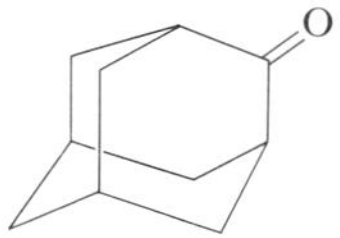 O	9808	—	2300	430	—	316	152	65
$Eu(fod)_3$	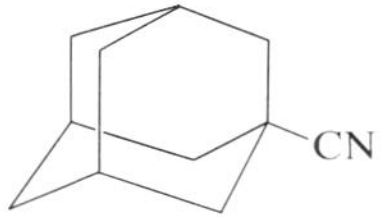 CN	—	—	480	185	—	22	45	65

[a] Heptane.
[b] Cyclohexane.

of analogous oxygen compounds,[104] but they show extremely weak binding to lanthanide shift reagents.[93,105–108]

The expected trends are observed for electron-donating and electron-withdrawing groups. The three fluorines of a trifluoroacetate, for example, result in very weak binding at the carbonyl oxygen of these esters.[42,88,109] Although the trifluoromethyl group is a powerful electron-withdrawing substituent, a single fluorine can behave in an unusual manner. San Filippo has reported[110] that substantial LISs can be observed with alkyl fluorides using $Yb(fod)_3$ as a shift reagent. On the other hand, he observed virtually no interaction with an aryl fluoride. Similarly, the LISs observed for polyhalogenated alkanes, such as difluoromethane and fluorodichloromethane, were extremely small.[110b] We have reported an upper limit of 10^{-3} for the association constant of $Eu(fod)_3$ with 1-bromoadamantane,[65] so coordination to halogen substituents appears to be unique to monofluoroalkyl groups. A review of x-ray structures of organic fluorine compounds has led to the conclusion that fluorine also can act as an electron donor in interactions with alkali-metal cations.[111]

Typical aromatic substituent effects are seen for the association constants of a series of substituted acetophenones[84] and of anilines.[71] Similar effects (presumably resulting from variations in the association constants) were reported by Mannschreck[112] for the corresponding LISs of substituted anilines, and the observed LISs of a series of substituted methyl benzoates[113] also reflect these substituent effects on association constants. Electron-withdrawing groups greatly reduce the association constants, whereas electron-donating groups result in stronger binding. (One must be careful, however, that an electron-donating group is not the actual coordination site in a polyfunctional compound.)

Steric effects also can play an important role in determining the magnitude of the association constants for shift reagents with organic substrates, although the influence is less than for electronic effects. Steric effects are greatest for amines, where alkyl substitution can be close to the actual coordination site or even on the donor atom. For other functional groups, such as alcohols or ketones, substitution must occur at the carbon atom to which the donor atom is bonded or at a position that is even further from the binding site. Clearly the steric interactions is larger when alkyl substituents are in close proximity to the binding site (and hence to the ligands of the shift reagent). The effect of alkyl substitution on an amine is illustrated by the large reduction in the association constant (with $Eu(dpm)_3$ in deuteriochloroform) form **15**–**17**.[73c]

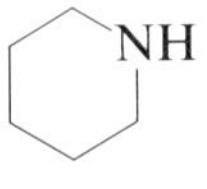

15 — K = 1360

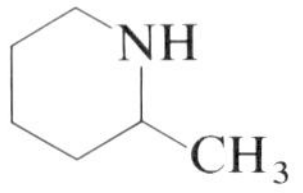

16 — K = 314

17 — K = 17

The association constants for pyridine (**10**) and its 2-methyl derivative (**18**) also illustrate the magnitude of steric effects for amines, whereas methyl substitution of the analogous nitriles actually results in a small increase in the association constant (presumably as a consequence of the electron-donating effect of the methyl group).

10	**18**	**19**	**20**
K = 333	K = 83	K = 19	K = 51
$Eu(dpm)_3/CDCl_3$		$Eu(fod)_3/CCl_4$	

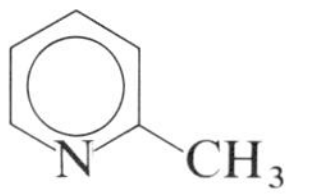

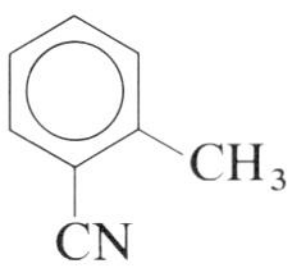

Two other variables that should be considered when attempting to estimate an association constant are the shift reagent and the solvent. The most commonly used shift reagents have been europium and ytterbium derivatives, in large part because they induce relatively large shifts without unacceptable line broadening. The line broadening is greater for ytterbium, but the magnitude of its induced shifts are greater than those of the europium reagents by a factor of about 3. Praeseodymium reagents have also been used by a number of workers because upfield shifts are induced (in contrast to the downfield shifts observed with europium and ytterbium derivatives). Relatively few data are available regarding association constants for different shift reagents. The most extensive comparison available involves the results of two independent studies of a series of substituted cyclohexanols,[31,114] and the results are summarized in Table 3-6.

The data in Table 3-6 (togther with various entries in Table 3-5) clearly demonstrate that $(fod)_3$ reagents are stronger Lewis acids than the dpm analogs. The fod reagents appear to bind more strongly by about a factor of 10 (after estimating the effect of solvent), and the fod/$(dpm)_3$ ratio varies directly with the binding ability of substrate for the same solvent and ligand. Europium reagents appear to have association constants that are larger than those for the corresponding ytterbium derivatives by about a factor of 10. This difference is surprising in light of the opposite behavior observed for complexation of lanthanide ions in aqueous solution.[8] A trend of increasing association constants might have been expected with increasing atomic number of the lanthanide.[124]

Also shown in Table 3-6 are the bound shifts for representative protons, and the data indicate that the same bound shifts are found for $Eu(dpm)_3$ and $Eu(fod)_3$. The relatively small discrepancies between the two reagents may reflect the different methods used to determine the bound shifts. In addition, the discrepancies may result in part from the greater association constants with $Eu(fod)_3$. We have found[61,66] previously that systematic errors in the association constant can result in corresponding errors in the bound shifts. However, the relative error in the bound shift is identical for all nuclei in the molecule, so this normally does not result in any significant problems.

The magnitude of solvent effects has not been investigated extensively, but we studied several compounds with the two-step method for a range of typical NMR solvents,[65] and a similar study was carried out by Bouquant and Chuche[115] with the equimolar method (Table 3-7). Significant variation in the association constants was observed as a function of solvent polarity. Solvent effects on the bound shifts were found to be small,[65,115] and the differences again may be only a consequence of the different mathematical procedures used to determine the bound shifts and association constants.

Despite their superiority in affording accurate bound shifts (and in yielding association constants that are otherwise unavailable), we must warn the reader of experimental difficulties with both the two-step and equimolar methods. The major difficulty with the two-step method lies in the complexity of the experiment. In order that both the 1:1 and 1:2 complexes be well determined, a large number of spectra must be obtained over a wide range of lanthanide–substrate concentration ratios. In our own work we ordinarily have obtained a total of 25 spectra for each experiment. As a consequence, sample preparation, recording the spectra, and analyzing the spectra are tedious operations. Nevertheless, no other method is available that affords bound shifts and association constants for both the 1:1 and 1:2 complexes and the more recent, simplified method[83] greatly reduces the difficulty.

The equimolar method is experimentally much more convenient.[61,70] Once the sample has been prepared, dilution with solvent is the only way that the sample can be changed further. Computations are also easier because only linear regression is needed. However, the magnitude of the shifts can be a major problem. Inspection of Table 3-3 shows that the bound fractions remain within a rather narrow range (eg, 0.17–0.50 for compound D), and this means that the observed shifts are always quite different from those in the absence of shift reagent. Although the changes in observed chemical shift usually can be followed easily, there may be some difficulty in correlating the signals with those in the unshifted spectrum. This difficulty can be resolved fairly easily, however, by carrying out a qualitative study in which shift reagent is added to a solution of substrate until a molar ratio of about 1:1 is reached. The spectra obtained in this way allow correlation of signals in the unshifted and shifted spectra.

Accurate measurement of the concentration following dilution with solvent can create difficulties with the equimolar method. The quantity of added solvent can be determined very accurately by weighing the sample tube before and after dilution. However, conversion of the change in weight to a change in volume requires the assumption that the volumes of the solvent increment and the original sample solution are strictly additive. Alternatively, the height of the liquid in the NMR tube can be measured and converted to volume. (The NMR tube can be calibrated easily for height vs. volume by weighing increments of pure solvent with known density.) We have used both approaches, and both afford the same results within experimental error.[61,93]

Another potential difficulty with the equimolar method is the requirement that shift reagent and substrate have identical concentrations. Our previous study[61] demonstrated that as long as the lanthanide–substrate ratio was between 0.99 and 1.01, the resulting errors in bound shifts and association constants were not more than 1%. Because the necessary concentration ratio can be achieved with confidence, preparation of the initial sample does not pose a major obstacle.

Modified Gradient Method

Are there ever situations in which either the gradient or the reciprocal method would be the method of choice? With regard to the reciprocal method the answer is a reluctant no. The experimental procedure requires very low lanthanide–substrate ratios, and as a result only a small fraction of the substrate is ever complexed. This means that neither the 1:1 nor the 1:2 complex is well determined,[81,82] so despite the mathematical consistency of the method (within the context of assuming only a single complex), it does not appear to be capable of yielding correct LIS parameters. In addition, the low lanthanide–substrate ratios result in only very small induced shifts. Consequently, many overlapping resonances for a complex molecule may not be resolved, and it often is impossible to determine the chemical shifts for all the nuclei in the molecule.

The gradient method does offer some promise as a useful technique, however. In the first place, it is experimentally convenient—far more so than any of the other methods. Small increments of shift reagent can be added easily and accurately, especially if the shift reagent is added in the form of a relatively concentrated solution in the same solvent that is already being used for the NMR experiment. Such additions cause only minor changes in the total volume of the solution as long as a low lanthanide–substrate ratio is maintained. Because the gradient plot is usually linear only below a value of 0.4 for L_O/S_O, large amounts of shift reagent are not added to the sample. Sample preparation is more difficult for both the equimolar and the two-step methods, because solutions must be prepared accurately at specific concentrations. In contrast, the specific concentration of substrate is not critical with the gradient method, as long as the ratios of lanthanide and substrate concentrations are determined accurately.

The major limitation—and it is a serious one—of the gradient method is its failure to afford the correct bound shift. In many cases, however, there may be a way to circumvent this difficulty. ApSimon showed in 1976[116] that the relative LISs for the various nuclei in a molecule are concentration independent under either of two circumstances: first, when only a single complex is formed, and second when each nucleus in the molecule exhibits the same ratio of bound shifts for the 1:1 and 1:2 complexes. Because bound shifts for the 1:2 complex usually are approximately 50–60% as large as those for the 1:1 complex, the second possibility is often a reasonable approximation. In 1978 we reported that a plot of relative shifts vs. the lanthanide–substrate ratio could be used to assess reliably the validity of that assumption.[38] If the ratio of the LS and LS_2 bound shifts is the same for all nuclei, then plots of relative LIS vs. the lanthanide–substrate ratio will be straight lines with slopes of zero. In other words, the relative shifts will be concentration independent. Deviations from zero slope (such as illustrated in Figure 3-2) indicate that the bound shifts for LS and LS_2 must not all have the same ratio.

In order to verify the concentration independence of the relative shifts, it is necessary to employ lanthanide–substrate ratios greater than unity. This provides an additional advantage, because with an association constant (K_1) of at least 50 M^{-1}, more than 80% complexation is achieved at a lanthanide–substrate ratio of 3. Assuming that the relative LIS have been shown to be concentration independent, the modified gradient method affords several favorable features:

1. Accurate relative bound shifts.
2. A reasonable estimate of the actual bound shift of the 1:1 complex from the observed LIS at high lanthanide–substrate ratios.
3. Experimental convenience in sample preparation.
4. Experimental convenience in analyzing spectra, because those spectra determined at high lanthanide–substrate ratios have fewer overlapping resonances.
5. Computational convenience, because only linear regression (over the region of L_O/S_O from 0.0 to 0.4) is needed. This capability is found on many pocket calculators.

Although the modified gradient method has not yet been explored fully, we feel that its convenience frequently makes it the method of choice for semiquantitative LIS analysis. When this approach is applicable (ie, when the relative LISs are constant) the only drawback should be the possibility that all of the bound shifts are in error by some unknown factor. The factor would be identical for all nuclei in the molecule, however, and this could be resolved by the use of an adjustable scaling factor during any attempted fitting of the LIS with a proposed structure. A similar type of scaling error often is observed for bound shifts determined by the two-step method (although the discrepancy is usually less than 5%), and we have shown that this is corrected readily during the structure-fitting process.[66] The convenience of the modified gradient method should prove particularly useful to organic chemists who need to obtain the maximum amount of structural information for a minimum investment of effort in spectroscopy and computing.

^{13}C Bound Shifts

Studies of ^{13}C LIS have not been as thorough as those for proton LIS, and there seem to be two major reasons for this. In the first place, contact shifts generally are expected to be more important for carbon as a consequence of increased electron density near its nucleus. Second, experimental difficulty is much greater for carbon spectra. Acquisition times are frequently on the order of an hour or more, so a shift reagent experiment that requires spectra at 25 different concentrations clearly is unsatisfactory. Even 8 or 10 spectra are inconvenient when sample preparation time is included. In order to maximize the amount of structural information obtained from the shift reagent experiment, it is desirable to obtain both carbon and proton spectra. Unfortunately, this is not possible in many situations, because two different instruments often are used for proton and carbon NMR. The instruments may use different size sample tubes, and they may have different operating temperatures. The latter problem is particularly severe, because both the shift reagent equilibria and the actual LISs are temperature dependent.[53–59] These problems are resolved for spectrometers with multinuclear probes, but how can the problem of acquisition time be circumvented?

One approach that we have investigated in our laboratory takes advantage of the fact that carbon and proton spectra are obtained with a single spectrometer on the same sample. Using the two-step method, the normal number of proton spectra is obtained, but carbon spectra are recorded for only a few different concentrations. The proton data serve to define the association constants, and nonlinear regression analysis of all shifts together affords accurate bound shifts for both proton and carbon. Because we have shown[83] recently that satisfactory bound shifts and association constants can be obtained from as few as six or seven spectra, the overall experiment can be simplified even further.[125] Similar approaches to obtaining ^{13}C bound shifts should also be satisfactory for the equimolar and gradient methods in the situations where those methods are applicable.

Polyfunctional Compounds

There are three methods by which polyfunctional compounds may be analyzed in shift reagent studies.

1. A complete mathematical analysis can be carried out for all the observed LIS as a function of concentration.
2. The use of appropriate blocking groups can be used to convert a difunctional compound to one that is effectively monofunctional.
3. If two functional groups are in close proximity, cooperative binding can result in behavior corresponding to that of a monofunctional compound.

The first alternative would be analogous to the two-step method, but far more equilibria would be involved. Coordination could occur at each binding site, so complexes would be expected consisting of a single substrate with more than one lanthanide. Similarly, isomeric LS_2 complexes would be possible because the two substrates might be bound at either the same or different sites. Considering the soft fit for a monofunctional substrate with the two-step method, a procedure for rigorous analysis of polyfunctional compounds does not appear feasible at the present time.

The second approach is one that we have explored in our laboratory.[88] The *t*-butyldimethylsilyl ether is uniformly effective at blocking hydroxy groups, and with the exception of primary hydroxy groups, the trimethylsilyl derivatives also work well. In contrast to earlier suggestions by us[42] and others,[6,109,117,118] trifluoroacetates are not adequate as blocking groups for rigorous shift reagent studies. Although trifluoroacetates bind only weakly to shift reagents, coordination at that site can be responsible for a significant contribution to the induced shifts of nuclei that are in close proximity.[88] Consequently, the silyl ethers are superior as blocking groups, and their use has permitted rigorous structure analysis with polyfunctional compounds such as the steroid derivatives **21** and **22**.

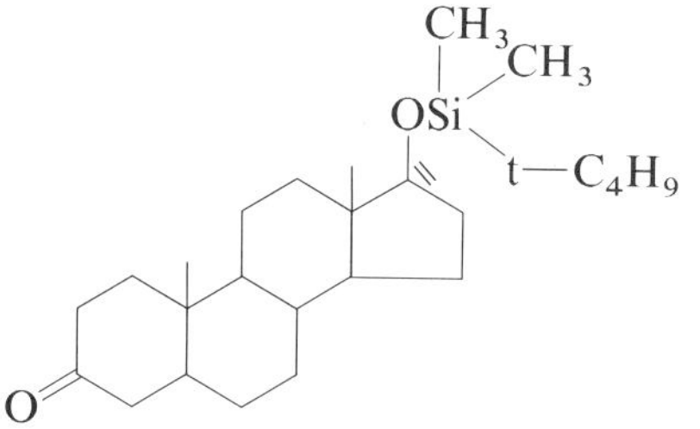

CH_3 CH_3 OSi t—C_4H_9 O

21

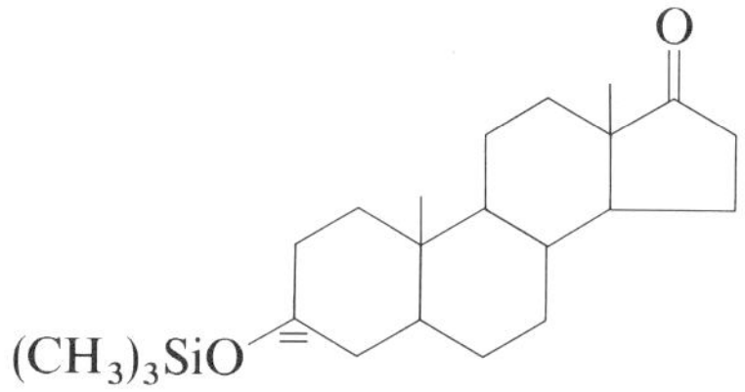

O $(CH_3)_3SiO$

22

The third way that polyfunctional compounds can be analyzed with shift reagents is also one that we have investigated.[39] For compounds with two functional groups in close proximity, binding via a five- or six-membered chelate ring appears to be quite favorable. Even when neither functional group by itself binds strongly, cooperative binding is observed for the difunctional compound. This can be seen for the interaction of $Eu(fod)_3$ with the fluorenones **23** and **24**. The electron-withdrawing effect of the aromatic rings results in rather weak binding for the parent ketone **23**, and aryl ethers have long been known to bind only very weakly to shift reagents.[36,73f,99] Nevertheless, the association constant for the methoxyfluorenone **24** is between one and two orders of magnitude larger than that of **23**.

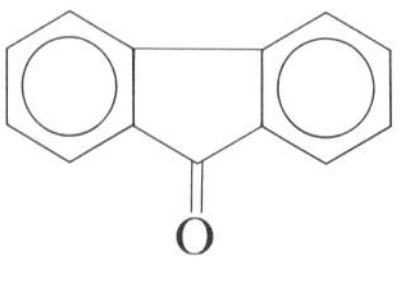

O

23

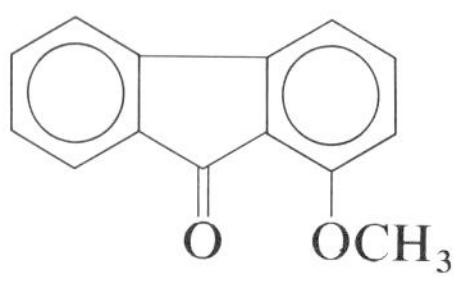

O OCH_3

24

As illustrated by the polyfunctional compounds in Table 3-5, increased binding strength as a consequence of chelation is well precedented. It is supported further by other studies in which the magnitude of the lanthanide-induced shifts indicates cooperative binding by two functional groups in a substrate molecule.[119–122] A series of oxygen heterocycles related to **23** and **24** was studied by Willcott,[123] and he similarly concluded that chelation was important when it was geometrically possible. A significant feature of the interaction with such chelating is the formation of an 8-coordinate lanthanide by what is effectively a one-step mechanism. Consequently, the equimolar method would be the procedure of choice for analysis. Even for compounds with much larger association constants this would be true, because there still would be only a single complex in solution.

Conclusions

Lanthanide shift reagents have an unusual history. Their initial discovery as aids to NMR spectroscopy in 1969[1] was followed by a period of intense activity, and the promise of shift reagents as a method for solution phase-structure analysis seemed unlimited. However, the complexity of the shift reagent equilibria and of the magnetic interactions imposed considerable constraints on progress in the field. Simple model compounds could be analyzed with great success, but the more complex molecules of interest to a practicing chemist resisted rigorous analysis. Even the simple model structures could be studied rigorously only at the expense of considerable time and effort. Clearly this was not a method for routine use by the practicing organic chemist. As a consequence, interest in shift reagents as an aid to rigorous structure determination declined substantially after the initial surge. Shift reagents have been used routinely for spectral simplification and for some qualitative structure evaluation, but their use for quantitative structure work has been very limited.

Only in the last several years have there been indications that shift reagents may, after all, be of great value for critical structure evaluation. A limited number of research groups around the world have carried out careful studies to define the scope and limitations of the method. Certainly one of the most important problems has been to find a method for determining LIS that is both accurate and reliable. This review has shown that several alternatives are available, although one must be careful to select the best method for the problem at hand. Even more importantly, it is apparent that the time has arrived for some experimental simplifications to be made. Several procedures have been discussed here for the experimental determination of LIS that offer great reduction in experimental complexity with only minor costs in terms of experimental accuracy. Lanthanide shift reagents once again are offering a powerful structural tool for organic chemists. However, this time the improved experimental procedures are likely to allow LIS to receive much wider use than has previously been possible.

References

1. Hinckley, C. C. *J. Am. Chem. Soc.*, **1969**, *91*, 5160.
2. Inagaki, F.; Miyazawa, T. *Prog. Nucl. Magn. Reson. Spectrosc.*, **1981**, *14*, 67.
3. Hofer, O. In "Topics in Stereochemistry"; Allinger, N. L.; Eliel, E. L. Eds.; John Wiley: New York, 1976; Vol. 9, p. 111.
4. Reuben, J. *Prog. Nucl. Magn. Reson. Spectrosc.*, **1973**, *9*, 1.
5. Mayo, B. C. *Chem. Soc. Rev.*, **1973**, *2*, 49.
6. Cockerill, A. F.; Davies, G. L. O; Harden, R. C.; Rackham, D. M. *Chem. Rev.*, **1973**, *73*, 553.
7. Reuben, J.; Elgavish, G. A. In "Handbook on the Physics and Chemistry of Rare Earths"; Gschneider, K. A.; Eyring, L. Eds.; North-Holland: Amsterdam, 1979; p. 483.
8. Peters, J. A.; Kieboom, A. P. G. *Recl. Trav. Chim. Pays-Bas*, **1983**, *102*, 381.
9. Roth, K.; Grosse, M.; Rewicki, D. *Tetrahedron Lett.*, **1972**, 435.
10. Sanders, J. K. M.; Hanson, S. W.; Williams, D. H. *J. Am. Chem. Soc.*, **1972**, *94*, 5325.
11. Shapiro, B. L.; Johnston, M. D., Jr. *J. Am. Chem. Soc.*, **1972**, *94*, 8185.
12. Gibb, V. G.; Armitage, I. M.; Hall, L. D.; Marshall, A. G. *J. Am. Chem. Soc.*, **1972**, *94*, 8919.
13. ApSimon, J. W.; Beierbeck, H.; Fruchier, A. *J. Am. Chem. Soc.*, **1973**, *95*, 939.
14. Reuben, J. *J. Am. Chem. Soc.*, **1973**, *95*, 3534.
15. Cunningham, J. A.; Sands, D. E.; Wagner, W. F.; Richardson, M. F. *Inorg. Chem.*, **1969**, *8*, 22.
16. Richardson, M. F.; Corfield, P. W. R.; Sands, D. E.; Sievers, R. E. *Inorg. Chem.*, **1970**, *9*, 1632.
17. Erasmus, C. S.; Boeyens, J. C. A. *Acta Crystallogr., Sect. B*, **1970**, *26*, 1843.
18. Zalkin, A.; Templeton, D. H.; Karraker, D. G.; *Inorg. Chem.*, **1969**, *8*, 2680.
19. Cotton, F. A.; Legzdins, P. *Inorg. Chem.*, **1968**, *7*, 1777.
20. Watkins, E. D., II; Cunningham, J. A.; Phillips, T., II; Sands, D. E.; Wagner, W. F. *Inorg. Chem.*, **1969**, *8*, 29.
21. Phillips, T., II; Sands, D. E.; Wagner, W. F. *Inorg. Chem.*, **1968**, *7*, 2295.
22. de Velliers, J. P. R.; Boeynes, J. C. A. *Acta Crystallogr., Sect. B*, **1971**, *27*, 692.
23. Watson, W. H.; Williams, R. J.; Stemple, N. R. *Inorg. Nucl. Chem.*, **1972**, *34*, 501.
24. Cramer, R. E.; and Seff, K. *Acta Crystallogr., Sect. B*, **1972**, *28*, 3281.
25. Horrocks, W. DeW., Jr.; Sipe, J. P.; III; Luber, J. R. *J. Am. Chem. Soc.*, **1971**, *93*, 5258.
26. Ilinskii, A. L; Aslanov, L. A.; Ivanov, V. I.; Khalilov, A. D. and Petrukhin, O. M. *Zh. Struct. Khem.*, **1969**, *10*, 285; *J. Struct. Chem.*, **1969**, *10*, 263.
27. Aslanov, L. A.; Korytnii; E. F., and Porai-Koshits, M. A. *Zh. Strukt. Khem.*, **1971**, *12*, 661; *J. Struct. Chem.*, **1971**, *12*, 600.

28. Lind, M. D.; Lee, B.; Hoard, J. L.; *J. Am. Chem. Soc.*, **1965**, *87*, 1611.
29. Hoard, J. L.; Lee, B.; Lind, M. D. *J. Am. Chem. Soc.*, **1965**, *87*, 1613.
30. Evans, D. F.; Wyatt, M. *Chem. Commun.*, **1972**, 312.
31. Johnston, M. D., Jr.; Shapiro, B. L.; Shapiro, M. J.; Proulx, T. W.; Godwin, A. D.; Pearce, H. L. *J. Am. Chem. Soc.*, **1975**, *97*, 542.
32. Feibush, B.; Richardson, M. F.; Sievers, R. E.; Springer, C. S., Jr. *J. Am. Chem. Soc.*, **1972**, *94*, 6717.
33. Archer, M. K.; Fell, D. S.; Jotham, R. W. *Inorg. Nucl. Chem. Lett.*, **1971**, *7*, 1135.
34. Desreux, J. F.; Fox, L. E.; Reilley, C. N. *Anal. Chem.*, **1972**, *44*, 2217.
35. Porter, R.; Marks, T. J.; Shriver, D. F. *J. Am. Chem. Soc.*, **1973**, *95*, 3548.
36. de Boer, J. W. M.; Hilbers, C. W.; de Boer, E. *J. Magn. Reson.*, **1977**, *25*, 437.
37. Volka, K.; Suchánek, M.; Karhan, J.; Hájek, M. *J. Mol. Struct.*, **1978**, *46*, 329.
38. Raber, D. J.; Johnston, M. D., Jr.; Campbell, C. M.; Janks, C. M.; Sutton, P. *Org. Magn. Reson.*, **1978**, *11*, 323.
39. Raber, D. J., Propeck, G. J. Unpublished results.
40. Cunningham, J. A.; Sievers, R. E. *Inorg. Chem.*, **1980**, *19*, 595.
41. Cramer, R. E.; Maynard, R. B.; Dubois, R. *J. Chem. Soc., Dalton Trans.*, **1979**, 1350.
42. Raber, D. J.; Johnston, M. D., Jr.; Campbell, C. M.; Guida, A.; Jackson, G. F., III; Janks, C. M.; Perry, J. W.; Propeck, G. J.; Raber, N. K.; Schwalke, M. A.; Sutton, P. *Monatsh. Chem.*, **1980**, *111*, 43.
43. Raber, D. J.; Johnston, M. D.; Jr. *Spectrosc. Lett.*, **1982**, *15*, 287.
44. Raber, D. J.; Beaumont W. E.; Johnston, M. D., Jr. *Spectrosc. Lett.*, **1982**, *15*, 329.
45. Evans, D. F.; Wyatt, M. *J. Chem. Soc., Dalton Trans.*, **1974**, 765.
46. Cramer, R. E.; Dubois, R. *J. Am. Chem. Soc.*, **1973**, *95*, 3801.
47. Cramer, R. E.; Dubois, R.; Seff, K. *J. Am. Chem. Soc.*, **1974**, *96*, 4125.
48. Bovee, W. M. M. J.; Alberts, J. H.; Peters, J. A.; Smidt, J. *J. Am. Chem. Soc.*, *104*, **1982**, 1632.
49. Peters, J. A.; van Bekkum, H.; Bovée, W. M. M. J. *Tetrahedron*, **1982**, *38*, 331.
50. Grotens, A. M.; Backus, J. J. M.; Pijpers, F. W.; de Boer, E. *Tetrahedron Lett.*, **1973**, 1467.
51. Lindoy, L. F.; Lip, H. C.; Louie, H. W.; Drew M. G. B.; Hudson, M. J. *J. Chem. Soc., Chem. Commun.*, **1977**, 788.
52. Lindoy, L. F.; Louie, H. W. *J. Am. Chem. Soc.*, **1979**, *101*, 841.
52a. We have recently succeeded in obtaining slow-exchange spectra of 2-adamantanone with $Yb(fod)_3$, but a temperature of -113°C was required. See ref. 125.
53. Bleaney, B. *J. Magn. Reson.*, **1972**, *8*, 91.
54. McGarvey, B. R. *J. Magn. Reson.*, **1979**, *33*, 445.
55. Golding R. M.; Pyykkö, P. *Mol. Phys.*, **1973**, *26*, 1389.
56. Sherry, A. D.; Yang, P. P.; Morgan, L. O. *J. Am. Chem. Soc.*, **1980**, *102*, 5755.
57. Cheng, H. N.; Gutowsky, H. S. *J. Phys. Chem.*, **1978**, *82*, 914.
58. Elgavish, G. A.; Reuben, J. *J. Magn. Reson.*, **1974**, *16*, 360.
59. Grotens, A. M.; Backus, J. J. M.; de Boer, E. *Tetrahedron Lett.*, **1973**, 1465.
60. For an earlier discussion of shift reagent equilibria and algorithms for obtaining bound shifts, see: Reuben, J. In "Nuclear Magnetic Resonance Shift Reagents"; Sievers, R. E., Ed.; Academic Press: New York, 1973; p. 341.
61. Raber, D. J.; Hardee, L. E. *Org. Magn. Reson.*, **1983**, *20*, 125.
62. Cawley, J. J.; Petrocine, D. V. *Org. Magn. Reson.*, **1974**, *6*, 544.
63. Sanders, J. K. M.; Williams, D. H. *J. Am. Chem. Soc.*, **1971**, *93*, 641.
64. Demarco, P. V.; Elzey, T. K.; Lewis, R. B.; Wenkert, E. *J. Am. Chem. Soc.*, **1970**, *92*, 5734.
65. Raber, D. J.; Johnston, M. D., Jr.; Janks, C. M.; Perry, J. W.; Jackson, G. F., III. *Org. Magn. Reson.*, **1980**, *14*, 32.

66. Raber, D. J.; Janks, C. M.; Johnston, M. D., Jr.; Raber, N. K. *Org. Magn. Reson.*, **1981**, *15*, 57.
67. (a) Armitage, I.; Dunsmore, G.; Hall, L. D.; Marshall, A. G. *Chem. Commun.*, **1971**, 1281; (b) *Can. J. Chem.*, *50*, 2119 (1972).
68. Armitage, I. M.; Hall, L. D.; Marshall, A. G.; Werbelow, L. G. In "Nuclear Magnetic Resonance Shift Reagents"; Sievers, R. E., Ed.; Academic Press; New York, 1973; p. 313.
69. Kelsey, D. R. *J. Am. Chem. Soc.*, **1972**, *94*, 1764.
70a. Bouquant, J.; Chuche, J. *Tetrahedron Lett.*, **1972**, 2337; (b) *Bull. Soc. Chim. Fr.*, **1977**, 959.
71. Rackham, D. M.; Cockerill, A. F. *Spectrosc. Lett.*, **1977**, *10*, 305.
72. Rackham, D. M.; Binks, J. *Spectrosc. Lett.*, **1979**, *12*, 17.
73. (a) Rackham, D. M.; *Org. Magn. Reson.*, **1979**, *12*, 388; (b) *Spectrosc. Lett.*, **1979**, *12*, 603; (c) **1980**, *13*, 509; (d) **1980**, *13*, 513; (e) **1980**, *13*, 517; (f) **1981**, *14*, 117.
74. Raber, D. J.; Johnston, M. D., Jr.; Perry, J. W.; Jackson, G. F., III. *J. Org. Chem.*, **1978**, *43*, 229.
75. Raber, D. J.; Johnston, M. D., Jr.; Schwalke, M. A. *J. Am. Chem. Soc.*, *99*, **1977**, 7671.
76. Johnston, M. D., Jr.; Raber, D. J.; DeGennaro, N. K.; D'Angelo, A.; Perry, J. W. *J. Am. Chem. Soc.*, **1976**, *98*, 6042.
77. Raber, D. J.; Janks, C. M.; Johnston, M. D., Jr.; Schwalke, M. A.; Shapiro, B. L.; Behelfer, G. L. *J. Org. Chem.*, **1981**, *46*, 2528.
78. Raber, D. J.; Caines, G. H.; Johnston, M. D., Jr.; Raber, N. K. *J. Magn. Reson.*, **1982**, *47*, 38.
79. Inagaki, F.; Takahashi, S.; Tasumi, M.; Miyazawa, T. *Bull. Chem. Soc. Jpn.*, **1975**, *48*, 853.
80. Inagaki, F.; Tasumi, M.; Miyazawa, T. *Bull. Chem. Soc. Jpn.*, **1975**, *48*, 1427.
81. Deranleau, D. A. *J. Am. Chem. Soc.*, **1969**, *91*, 4044; **1969**, *91*, 4050.
82. Lenkinski, R. E., Elgavish, G. A.; Reuben, J. *J. Magn. Reson.*, **1978**, *32*, 367.
83. Raber, D. J.; Peters, J. A. *Magn. Reson. Chem.*, **1985**, *23*, 621.
84. Kawaki, H.; Takagi, T.; Fujiwara, H.; Sasaki, Y. *Chem. Pharm. Bull.*, **1981**, *29*, 2397.
85. Porter, G. B.; Simpson, J. *Spectrosc. Lett.*, **1981**, *14*, 755.
86. Catton, G. A.; Hart, F. A.; Moss, G. P. *J. Chem. Soc., Dalton Trans.*, **1976**, 208.
87. Bruder, A. H.; Tanny, S. R.; Rockefeller, H. A.; Springer, C. S., Jr. *Inorg. Chem.*, **1974**, *13*, 880.
88. Raber, D. J.; Propeck, G. J. *J. Org. Chem.*, **1982**, *47*, 3324.
89. Kawaki, H.; Fujiwara, H.; Sasaki, Y. *Chem. Pharm. Bull.*, **1978**, *26*, 2694.
90. Brittain, H. G.; *J. Chem. Soc., Dalton Trans.*, **1979**, 1187.
91. Grosse, M.; Roth, K.; Rewicki, D. *Org. Magn. Reson.*, **1977**, *10*, 115.
92. Huet, J.; Fabre, O.; Zimmermann, D. *Tetrahedron*, **1981**, *37*, 3739.
93. Unpublished results from the author's laboratory.
94. Graddon, D. P.; Muir, L. *J. Chem. Soc., Dalton Trans.*, **1981**, 2434.
95. Hirayama, M.; Owada, M. *Bull. Chem. Soc. Jpn.*, **1979**, *52*, 1786.
96. Hirayama, M.; Sasaki, Y. *Chem. Lett.*, **1982**, 195.
97. Graddon, D. P.; Muir, L.; Lindoy, L.; Louie, H. W. *J. Chem. Soc., Dalton Trans.*, **1981**, 2596.
98. Lindoy, L. F. *Coord. Chem. Rev.*, **1983**, *48*, 83.
99. Hofer, O. *Monatsh. Chem.*, **1978**, *109*, 405.
100. Lewin, A. H.; Alekel, R. *J. Am. Chem. Soc.*, **1976**, *98*, 6919.
101. Gibb, V. G.; Armitage, I. M.; Hall, L. D.; Marshall, A. G. *J. Am. Chem. Soc.*, **1972**, *94*, 8919.
102. Brittain, H. G.; Richardson, F. S. *J. Chem. Soc., Dalton Trans.*, **1976**, 2253.
103. Brittain, H. G. *J. Am. Chem. Soc.*, **1979**, *101*, 1733.

104. Wolf, J. F.; Staley, R. H.; Koppel, I.; Taagepera, M.; McIver, R. T., Jr.; Beauchamp, J. L.; Taft, R. W. *J. Am. Chem. Soc.*, **1977**, *99*, 5417.
105. Morrill, T. C.; Opitz, R. J.; Mozzer, R. *Tetrahedron Lett.*, **1973**, 3715.
106. Hájek, M.; Janků, J.; Burkhard, J.; Vodička, L. *Coll. Czech. Chem. Commun.*, **1976**, *41*, 2533.
107. van Bruijnsvoort, A.; Kruk, C.; de Waard, E. R.; Huisman, H. O. *Tetrahedron Lett.*, **1972**, 1737.
108. Hart, H.; Love, G. M. *Tetrahedron Lett.*, **1971**, 625.
109. Ius, A.; Vecchio, G.; Carrea, G. *Tetrahedron Lett.*, **1972**, 1543.
110. (a) San Filippo, J., Jr.; Nuzzo, R. G.; Romano, L. J. *J. Am. Chem. Soc.*, **1975**, *97*, 2546; (b) *J. Org. Chem.*, **1976**, *41*, 392.
111. Murray-Rust, Stallings, W. C.; Monti, C. T.; Preston, R. K.; Glusker, J. P. *J. Am. Chem. Soc.*, **1983**, *105*, 3206.
112. Ernst, L.; Mannschreck, A. *Tetrahedron Lett.*, **1971**, 3023.
113. Ranganayakulu, K.; Dubey, R.; Rajeswari, K. *Tetrahedron Lett.*, **1981**, *22*, 4359.
114. Bouquant, J.; Maugean, A.; Chuche, J. *Can. J. Chem.*, **1979**, *57*, 1080.
115. Bouquant, J.; Chuche, J. *Tetrahedron Lett.*, **1973**, 493.
116. ApSimon, J. W.; Beierbeck, H.; Fruchier, A. *Org. Magn. Reson.*, **1976**, *8*, 483.
117. Girard, P.; Kagan, H.; David, S. *Tetrahedron*, **1971**, *27*, 5911.
118. Crump, D. R.; Sanders, J. K. M.; Williams, D. H. *Tetrahedron Lett.*, **1970**, 4949.
119. Hofer, O.; Griengl, H.; Nowak, P. *Monatsh. Chem.*, **1978**, *109*, 21.
120. McArdle, P.; O'Reilly, J. P.; Simmie, J.; Lee, E. E. *Carbohydr. Res.*, **1981**, *90*, 165.
121. Camps, P.; Font, J.; Marques, J. M. *Tetrahedron*, **1975**, *31*, 2581.
122. Warrener, R. N.; Pitt, I. G.; Russell, R. A. *J. Chem. Soc., Chem. Commun.*, **1982**, 1195.
123. Davis, R. E.; Willcott, M. R., III. In "Nuclear Magnetic Resonance Shift Reagents"; Sievers, R. E., Ed.; Academic Press: New York, 1973; p. 143.
124. In a recent study we have found that the values of K_1 do not vary greatly for the interaction of $Ln(fod)_3$ with adamantanone in $CDCl_3$ (La = 60, Eu = 94, Gd = 74, Dy = 229, Yb = 82 M^{-1}): J. A. Peters and D. J. Raber, unpublished results.
125. J. A. Peters, M. S. Nieuwenhuizen and D. J. Raber, *J, Magn. Reson.*, **1985**, *65*, 417.

4

SOLUTIONS TO STEREOCHEMICAL PROBLEMS

Jaakko Paasivirta

DEPARTMENT OF CHEMISTRY,
UNIVERSITY OF JYVÄSKYLÄ,
FINLAND SEMINAARINKATU 15, SF-40100,
JYÄSKYLÄ 10, FINLAND

Introduction

In the other chapters of this volume the usefulness of certain lanthanide *tris*-diketonates in resolving NMR signals of organic compounds and the theory behind the phenomenon have been described. The spatially anisotropic and specific nature of the main effect allows its use in determining the molecular geometry of the substrate.

According to present theories, lanthanide-induced shift (LIS) effects consist of two parts: the Fermi contact contribution and the dipolar (pseudocontact) term.[1] The theory of Golding and Halton[2] deduces the contact shift as proportional to spin polarization at the NMR nucleus, which is dependent only on the central lanthanide ion. In many cases the Fermi contact shift is negligible; the shift effects are then purely dipolar in nature and so provide an excellent tool in stereostructure determinations. The recent theories of Bleaney[3] and Golding and Pyykkö[4] predict the signs, magnitudes, and temperature dependences of the pseudocontact effects for each lanthanide, and the results confirm that this effect

Morrill (ed): LANTHANIDE SHIFT REAGENTS IN STEREOCHEMICAL ANALYSIS

is proportional to the magnetic susceptibility anisotropy.[5] In the NMR sample solution the rapid exchange between free and complexed substrate and shift reagent results in a statistically averaged NMR spectrum of different substrate species. Consequently, the observed shift effect values in most cases obey the axially symmetrical pseudocontact equation of McConnell and Robertson:[5]

$$\mathrm{LIS} = \mathrm{k}\left(\frac{3\cos^2\theta - 1}{\mathrm{R}^3}\right) + \mathrm{b} \qquad (4\text{-}1)$$

where R is the distance between the NMR nucleus and the paramagnetic center (lanthanide ion), θ is the angle of the R vector, or paramagnetic center with the principal magnetic axis (in most cases the axis of the coordinative bond between the lanthanide and the *n* donor atom of the substrate), and LIS is the lanthanide-induced shift. This relationship provides an excellent opportunity to verify the stereostructure of the substrate by a quantitative treatment of the LIS values. As one of the first applications, Farid et al[6] designed a computer program to determine the location of the lanthanide ion in the complex with the organic molecule which gives the smallest sum of squared errors between the measured shifts and those predicted from eq. (4-1). This iterative procedure can be applied to determining the positions of the NMR nuclei in the substrate,[7] too, and so partly unknown stereostructures can also be deduced.

The above quantitative treatment of LIS data is in most practical cases only a completion of the structural information of a more qualitative nature obtained with the aid of the increased NMR resolution achieved by in the presence of shift reagents.[8] For example, when the signals of the individual skeleton protons become separate, integrated areas, signal multiplicities, including coupling constants and spin–spin decoupling experiments, lead to complete interpretation of the signals and to the verified structure of the substrate.[9]

Experimental Remarks

The first generally adopted method for measuring LIS values[10] consisted of preparing substrate–shift reagent solution of known concentrations in NMR solvent ($CDCl_3$, $CHCl_3$, or CCl_4), recording the spectrum, and then adding a weighed amount of shift reagent, recording the spectrum again, etc. The LIS values for different NMR nuclei were calculated as the slope of a plot of δ, which is the chemical shift for a given nucleus in the presence of a given increment of shift reagent, versus the ratio of the concentrations of shift reagent and substrate. The results are seriously confused by the influence of water or other impurities, by uncertainty in the absolute concentrations, and by dimerization of certain shift reagents.[11] Most of these drawbacks can be eliminated by using another procedure, the constant-S_O incremental dilution method, in which S_O is the initial concentration of substrate.[12,13] In this method the first experimental sample is prepared at the highest lanthanide shift reagent (LSR) concentration to be run, and subsequent samples are prepared by dilution with a substrate stock solution of the same substrate molarity as the first sample.

Hygroscopic reagents, such as $Eu(fod)_3$, have necessitated doing the NMR sample preparations in a dry box under a dry nitrogen atmosphere.[14]

In our laboratory the availability of the substrate has been the limiting factor in the choice of an NMR sampling method. In many cases the compound to be studied was collected from preparative gas chromatography (GC) in milligram amounts without knowing the exact weight. For these cases we developed an LSR addition method for determining relative LIS values:[15,16] a sample of the substrate was directed from a gas chromatograph into a solution of carbon tetrachloride. The ^{1}H NMR spectrum was run (after a little tetramethylsilane, TMS, was added) and again after a weighed amount of LSR was added. Then more LSR was weighed and added and the spectrum was run again, and so on. Thus a series of spectra from samples containing different LSR–substrate ratios was obtained from a single sample of substrate. For assignments of the ^{1}H NMR signals, integrals and spin decoupling experiments were run before the next LSR addition, whenever necessary. In most cases the dependence of the chemical shift of each proton on the added amount of LSR proved to be linear over a wide range, indicating the dominance of the equilibrium between free and 1:1 complexed substrate and LSR in solution. The shifts (in ppm) can be plotted against added milligrams of LSR and the slopes of the lines thus obtained can be used as relative LIS values. The LISs of some proton at a position of known stereochemistry can be used as an intramolecular standard.[9]

For ^{1}H NMR, choice of the LSR is not very critical because the pseudocontact effect dominates except for some protons very near to the coordination site.[17] In planning quantitative treatment of LISs the possibility of 1:2 LSR–substrate complexes in the solution must be considered.[12] However, these seem to have significance at low concentrations only for $Eu(fod)_3$.[18]

For ^{13}C NMR, especially when quantitative structural treatment is of interest, there are more significant contact contributions and these have to be considered when choosing the shift reagent. For a given substrate the distribution of an isotropic (contact) shift can be estimated from the effect of $Gd(fod)_3$ and then compared with different LSR effects.[19] Europium contact effects are among the highest and most disturbing in carbon resonance.[20] Line broadening prevents the use of some others, such as Dy and Ho. This leaves the Yb complex as very useful for dual proton–carbon NMR, where line broadening is not much worse than that from Eu, and the purely dipolar shift effects induced by $Yb(dpm)_3$ are *ca.* 250% larger than for $Eu(dpm)_3$. Consequently, the ytterbium complex is recommended[20] when proton and carbon NMR–LIS data are intended to be used jointly for stereostructure determinations.

Qualitative Approach

Rigid structures with one atom containing coordinating *n* electron pairs are excellent models to study LIS effects. Among them 2-norbornyl compounds are of special interest because the question of the nature of the 2-norbornyl ion has been central in the development of organic chemistry during this century.[21] Norbornane, with a bicyclo[2.2.1]heptane skeleton, is very common in natural terpenes. The 2-norbornyl alcohols have two possible configurations that were difficult to resolve by structure chemical methods until Finnish chemists, led by Toivonen, succeeded in proving, by chemical methods, the endo or exo structures of borneol and isoborneol[22,23] and of fenchols.[24–26]

Proton NMR spectra of *exo*- and *endo*-2-norbornanols (**1** and **2**, respectively) were complex and unresolvable until lanthanide shift reagents were used.[9] Only the signals of α-protons at position 2 had been interpreted for deducing the configuration.[27] Use of the pyridine solvent effect and decoupling might produce some more information for **1** but not for **2**.[9,28]

The influence of the added $Eu(dpm)_3$ to the spectra of 2-norbornanols in CCl_4 is illustrated in Figures 4-1 and 4-2. The different signals are

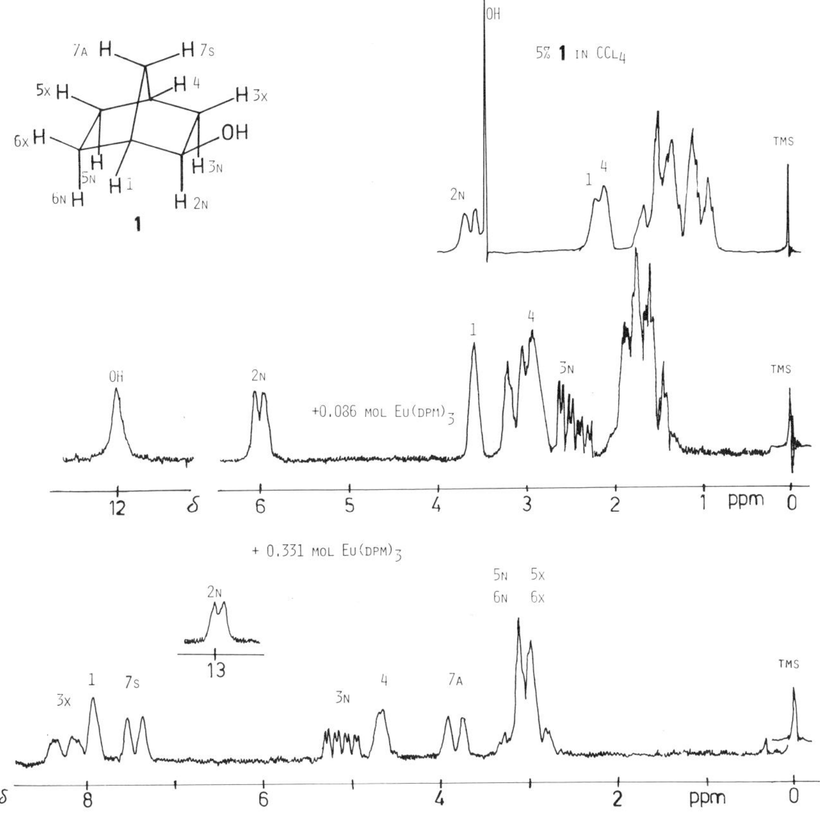

Figure 4-1. Proton NMR spectra of norbornan-2-*exo*-ol (**1**) in CCl_4 with and without added $Eu(dpm)_3$. Spectrometer: Varian A 60 A, 60 MHz, 35°C.

assigned readily by first-order analysis of their splitting patterns based on knowledge of the norbornane skeleton couplings obtained from **1** and from other model compounds with this skeleton.[28] Actually the spectra do not become first order and the small couplings are not resolved partly because of the line broadening. However, the large couplings can be read from the signals just by "first-order reasoning." Caution should be employed also in cases of "deceptive simplicity."[29] For example, the signal of the proton $2x$ of **2** (Figure 4-2) degenerates when the strongly coupled protons $3x$ and $3n$ by chance have the same chemical shifts (middle spectrum in Figure 4-2) but resolves again when more LSR is added (lower spectrum in Figure 4-2). Decision on the exo or endo configuration is best based on the large couplings of the α-protons at posi-

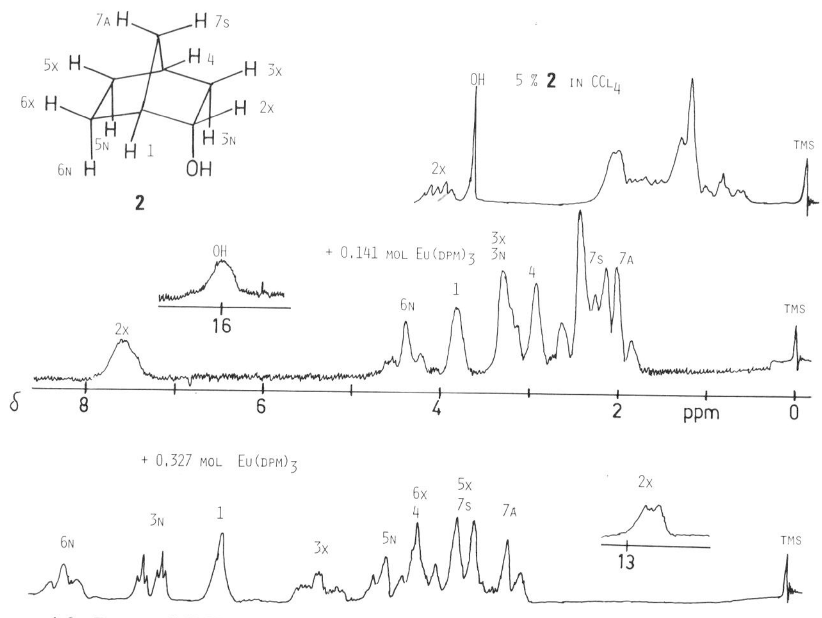

Figure 4-2. Proton NMR spectra of norbornan-2-*endo*-ol (**2**) in CCl_4 with and without added $Eu(dpm)_3$. Spectrometer: Varian A 60 A, 60 MHz, 35°C.

tion 2. Their approximative values are in Hz:

	Coupling constants					
Compound	2–3*x**	2–3*n*	2–1	2–7*a*	2–6*x*	3*x*–3*n*
exo-(**1**)	2.3	7.0	0	1.2	0	13.
endo-(**2**)	8.2	4.0	3.5	0	1.	13.

2-Norbornanols and the corresponding 5,6-unsaturated 2-norbornenols are also excellent models for studying the nature of the effects of shift reagents. Their NMR properties (couplings) are well known and coordinates of the atoms fixed; they also serve as good calibration means for quantitative treatments of LISs (see just below).

Experience with NMR[27,28] and LIS[9] studies of the secondary 2-norbornanols and 5-norbornen-2-ols (dehydronorborneols) is useful in structure determinations of tertiary 2-methyl analogs.[30–37] Spin decoupling experiments are performed for aid in signal assignments. For example, the spectrum of simple 2-*endo*-methyl-5-norbornen-2-*exo*-ol (**3**) can be interpreted in many parts by irradiating the proton signal at 2.8 ppm

* For example, compound **1** shows a coupling constant of 2.3 Hz between proton 2 and proton 3*x*.

(upper part of Figure 4-3) due to bridgehead hydrogen 4: the 3*x* signal degenerates to a doublet because $J_{3x\text{-}4}$ (4 Hz) is decoupled. After the spectrum of **3** is resolved by LSR addition, further spin decouplings can be done (lower parts of Figure 4-3). Irradiation of the bridge proton (7*a* signal, Figure 4-3a) shows the position of proton 7*s*. Irradiation of 4 shows, in addition to 3*x*, the olefinic proton 5 signal J_{45} (3 Hz) is decoupled (trace b in Figure 4-3). The double-resonance trace c (Figure 4-3)

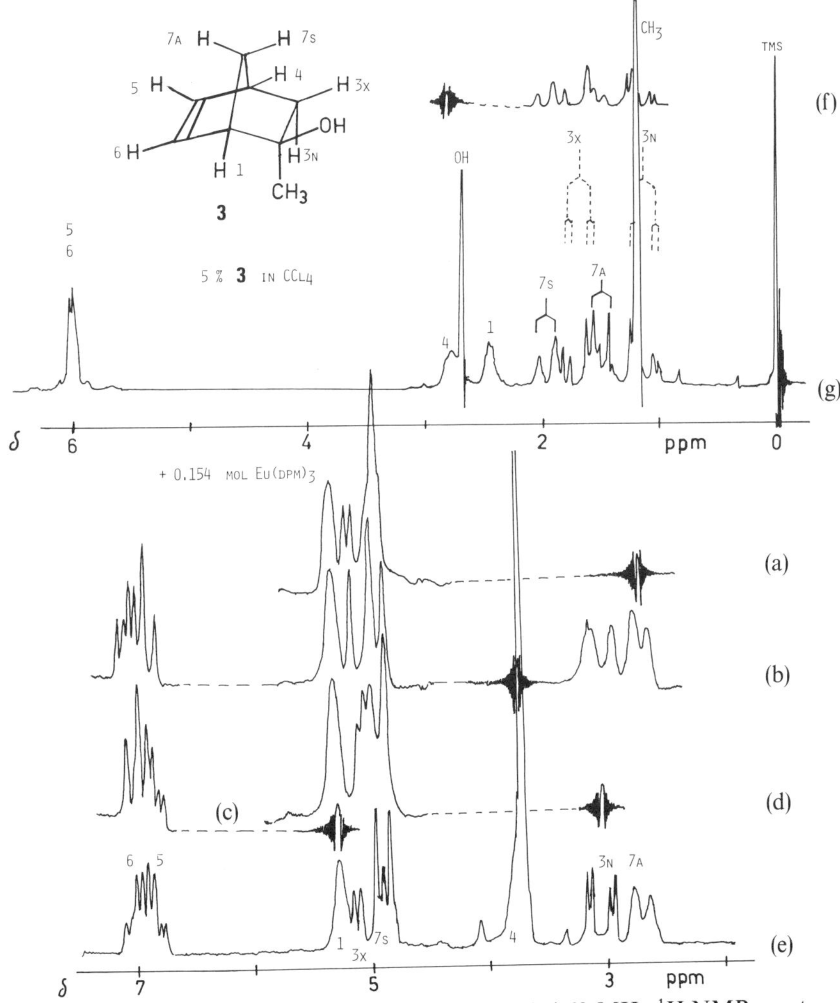

Figure 4-3. Nondecoupled and proton–proton decoupled 60-MHz ^{1}H NMR spectra of 2-*endo*-methyl-5-norbornen-2-*exo*-ol (**3**) in CCl_4 with and without added $Eu(dpm)_3$. Spectrometer: Perkin Elmer R 20 B. Parts (a)–(d) are LSR-shifted and partly decoupled. Part (e) is LSR-shifted only and part (f) is decoupled only. Part (g) is the unaltered spectrum.

shows proton 1 and 6 signals and trace d the decoupling between 3*n* and 3*x*. Similar experiments on 2-*exo*-methyl-5-norbornen-2-*endo*-ol (**4**, Figure 4-4) lead to complete assignments of the proton NMR signals in LIS-resolved spectra of **4**. Confirmation of the configurations can

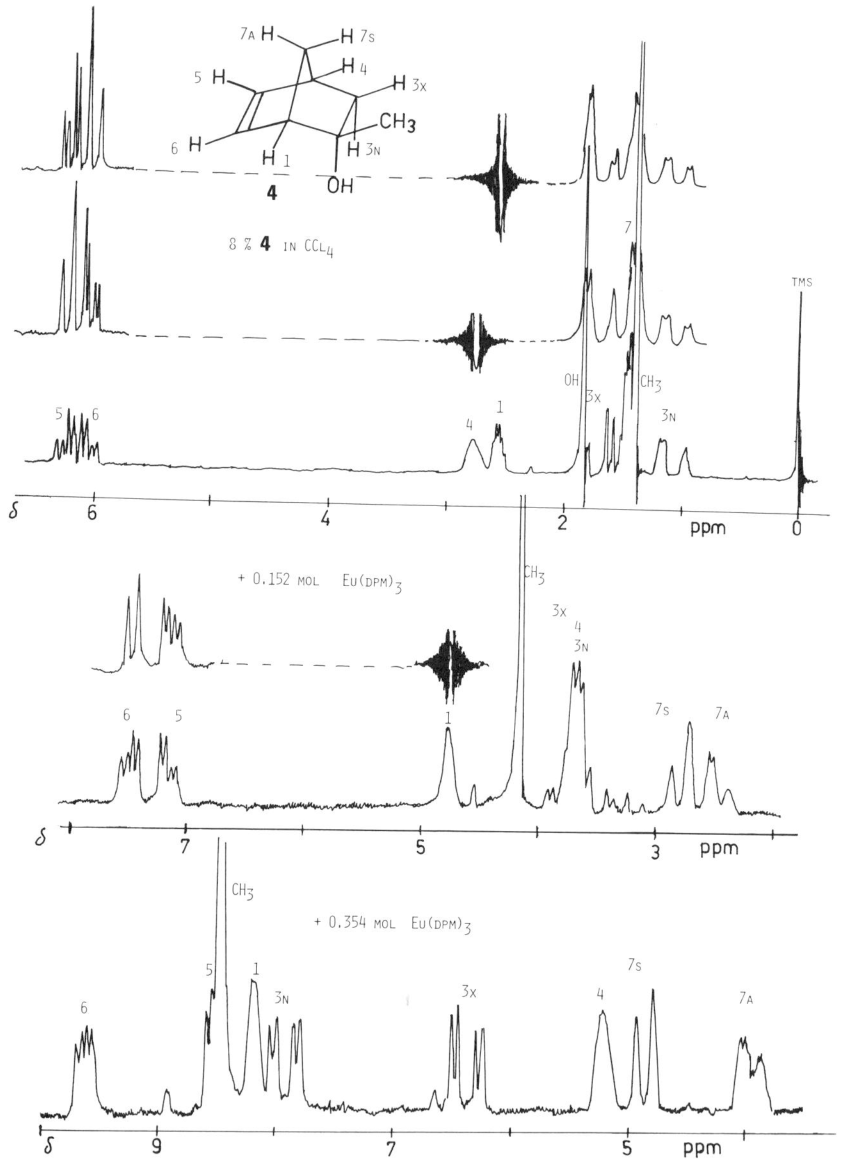

Figure 4-4. Nondecoupled and proton–proton decoupled ^1H NMR spectra of 2-*exo*-methyl-5-norbornen-2-*endo*-ol (**4**) in CCl_4 with and without added $Eu(dpm)_3$. Spectrometer: Perkin Elmer R 20 B.

be obtained by qualitative (or semiquantitative) evaluation of the LIS effects on each norbornene skeleton proton. The *exo*-OH structure for **3** is demonstrated by large shift effects on protons 3*x* and 7*s* and the *endo*-OH configuration of **4** by large shifts for 3*n* and 6. Experience with model compounds **3** and **4** was used for determining the structure of the main product of the Grignard reaction of dehydrocamphor with methyl iodide; this product proved to be 1,2-*endo*-7,7-tetramethyl-5-norbornen-2-*exo*-ol.[30]

The reaction of methyl Grignard with norcamphor gives the stereoisomeric alcohols **5** and **6** (Figures 4-5 and 4-6).[31,32] The major product was shown to be **6** by chemical methods using **3** and **4** as intermediates.[33–36] Proton NMR without LIS cannot distinguish between structures **5** and **6**. However, addition of $Eu(dpm)_3$ produces well-resolved spectra (Figures 4-5 and 4-6) and allows firm conclusions regarding their stereochemistry.[37] The *exo*-OH structure of the minor product **5** is verified by the large shift effects on 3*x* and 7*s* and the *endo*-OH configuration of the main product **6** by the large LIS effects on 3N and 6N.

Like the case of dehydrocamphor (see above), the product of the Grignard reaction of camphor had stereochemistry different from the main product of the norcamphor.[35,38,39] Its proton NMR could be further resolved by additions of $Eu(dpm)_3$ and the relative LIS values compared to those measured for borneol and isoborneol.[10] In this way the product structure was proved to be 1,2-*endo*-7,7-tetramethyl-2-*exo*-norbornanol or 2-methylisoborneol (MIB). In recent years this compound has become important as the cause of a bad taste of water and fish.[40,41] In nature, MIB is produced in slightly polluted water systems (also in reservoirs) as the metabolite of several actinomycetes (blue-green algae).

Qualitative and semiquantitative interpretation of LIS effects have been successful in proving the cis–trans stereoisomer structures of nortricyclanols. Examples are illustrated in Figure 4-7. Also, using 3-nortricyclanol (**7**) as a model, ^1H NMR LIS spectra of the two stereoisomeric 1-methyl-3-nortricyclanols (**8** and **9**) were interpreted with the aid of spin decoupling experiments.[42] The bridge methylene protons 5 and 7 are recognized from their typical geminal coupling (10 Hz) and the cyclopropane ring protons from their vicinal cis coupling (5–6 Hz). The LIS effects on protons 5 and 7 and on protons 2 and 6, however, cannot be used to differentiate stereoisomers **8** and **9**. In contrast, the methyl signals show a clear difference in LIS effects and give confirmation of a trans structure (**8**) for α and cis structure (**9**) for β-1-methyl-3-nortricyclanol; this was first proposed from GC behavior[43] and later shown by a series

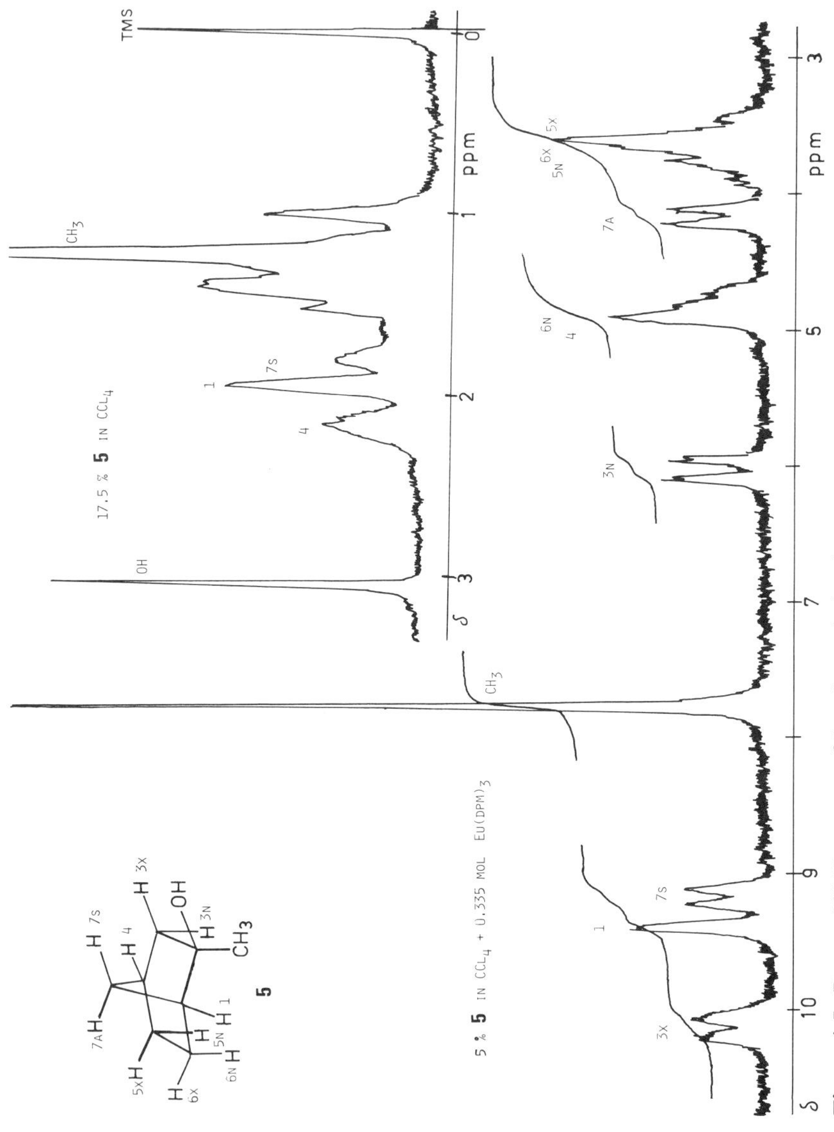

Figure 4-5. Proton NMR spectra of 2-*endo*-methylnorbornan-2-*exo*-ol (5) in CCl_4 before and after addition of 0.335 mol of $Eu(dpm)_3$; recorded at 90 MHz with a Bruker HFX 90 spectrometer.

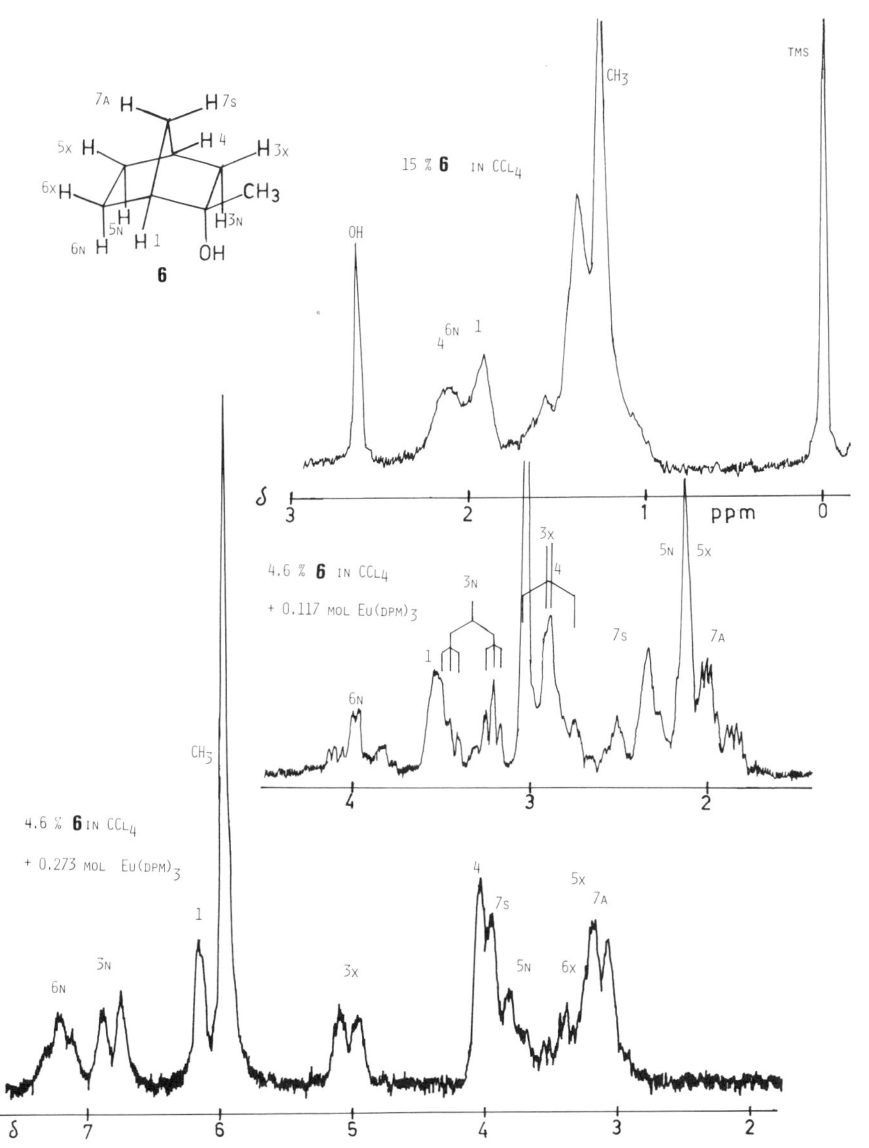

Figure 4-6. Proton NMR spectra of 2-*exo*-methylnorbornan-2-*endo*-ol (**6**) in CCl_4 before and after addition of 0.117 and 0.273 mol $Eu(dpm)_3$; recorded at 90 MHz with a Bruker HFX 90 spectrometer.

of syntheses and different structure determinations (including solvent effects on NMR).[44]

A qualitative investigation of the LIS effects on the ^1H NMR spectra of pericyclobornanol (**10**) is sufficient for determining its configuration

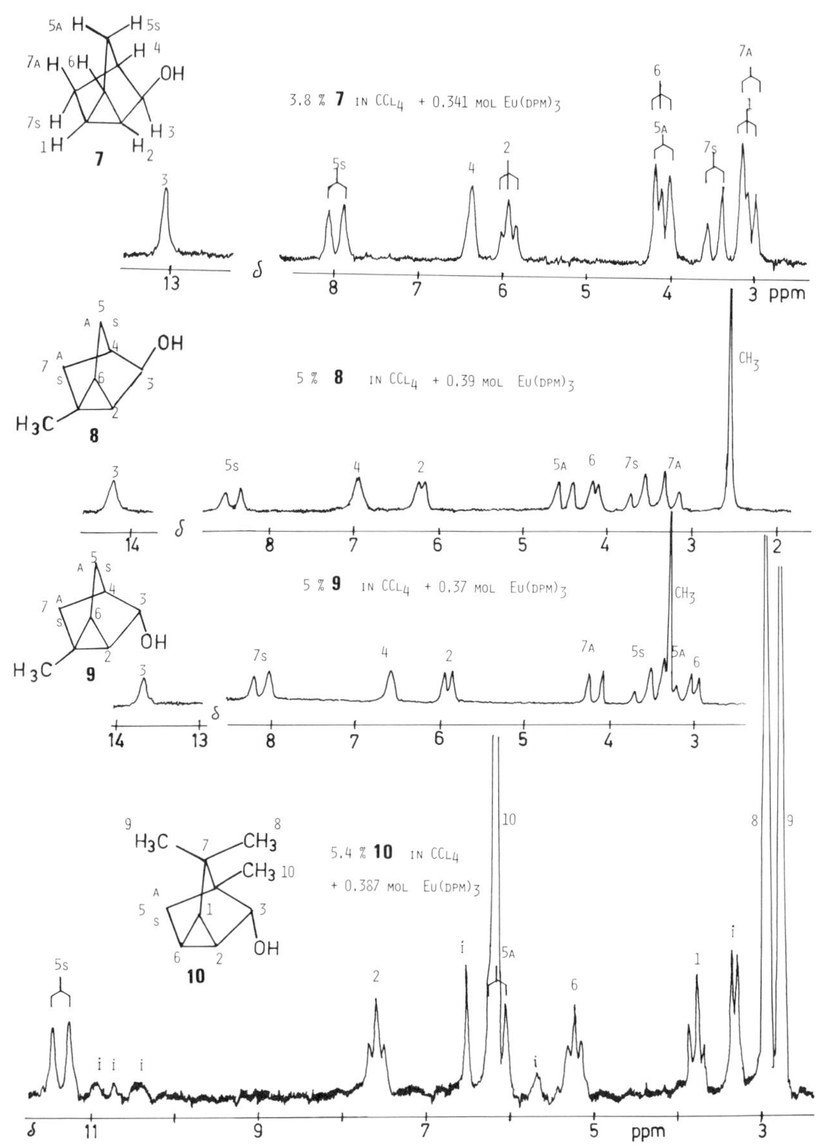

Figure 4-7. Proton NMR spectra of $Eu(dpm)_3$-containing CCl_4 solutions of 3-nortricyclanol (**7**), α-1-methyl-3-nortricyclanol (**8**), β-1-methyl-3-nortricyclanol (**9**), and pericyclobornanol (**10**); recorded with a Perkin Elmer R 12 B spectrometer at 60 MHz and 33.5°C.

(see lowest spectrum in Figure 4-7). Despite the fact that the sample was slightly impure, the signals could be assigned readily and the large effect noted of added $Eu(dpm)_3$ on one bridge proton signal (5*s*) and on only one methyl group signal (10). This is explained only if the configuration of the OH group is anti with respect to the methylated bridge 7 (structure **10**).

These qualitative deductions about the nortricyclanol structures have been verified by quantitative (computation) treatment of the LIS effects.[7]

Quantitative Approach

Using an iterative computer program[6,17] it is possible to optimize the lanthanide position with regard to the substrate molecule when an LSR–substrate complex is placed in a coordinate system.[45–47] As a result, the observed LIS values of the protons of the substrate are found to agree quite well with the calculated ones from eq. (4-1) if the conditions for the dominance of the pseudocontact effect and 1:1 complex formation of LSR and substrate are met (see introduction and experimental remarks section above). This is true when $Eu(dpm)_3$ is used in dilute solutions of CCl_4. Only a few protons near the coordination site may have significant contact contributions in their LIS values.[17] For results leading to the real structure, in most cases assignments of the protons must be correct.[14] If not, an incorrect minimum can be obtained in the computer fitting, which then gives deceptively small errors between observed and calculated LIS values. Such an error in minimum can be detected from observation of an unrealistic structure of the computed 1:1 complex, eg, from an unrealistically long or short lanthanide–heteroatom distance compared to the results with model compounds.

An illustrative example is obtained from the structure determination of Nojikigu alcohol, a camphene derivative from *Chrysanthemum* species. This compound was first detected by E. von Rudloff[48] in 1963 as "crystalline alcohol" from the volatile oil of *Tanacetum vulgare L.*, which is the same as *Chrysanthemum vulgare* (L.) Bernh. It was characterized in more detail by von Schantz and Forsen 1971[49] and given to our laboratory for NMR–LIS structure determination.[16] Simultaneously, Matsuo and his co-workers isolated the same compound from *Chrysanthemum japonese*, determined its structure by different spectroscopic methods, and gave to it the name Nojikigu alcohol.[50]

Spectra obtained by the method appropriate for an unknown amount collected from GC effluents[15,16] (see Experimental Remarks, above) of Nojikigu alcohol (**11**) are illustrated in Figure 4-8. Assignments of the signals were based on the integrals, chemical shifts, shift effects, and characteristic coupling distributions, which were verified later by spin

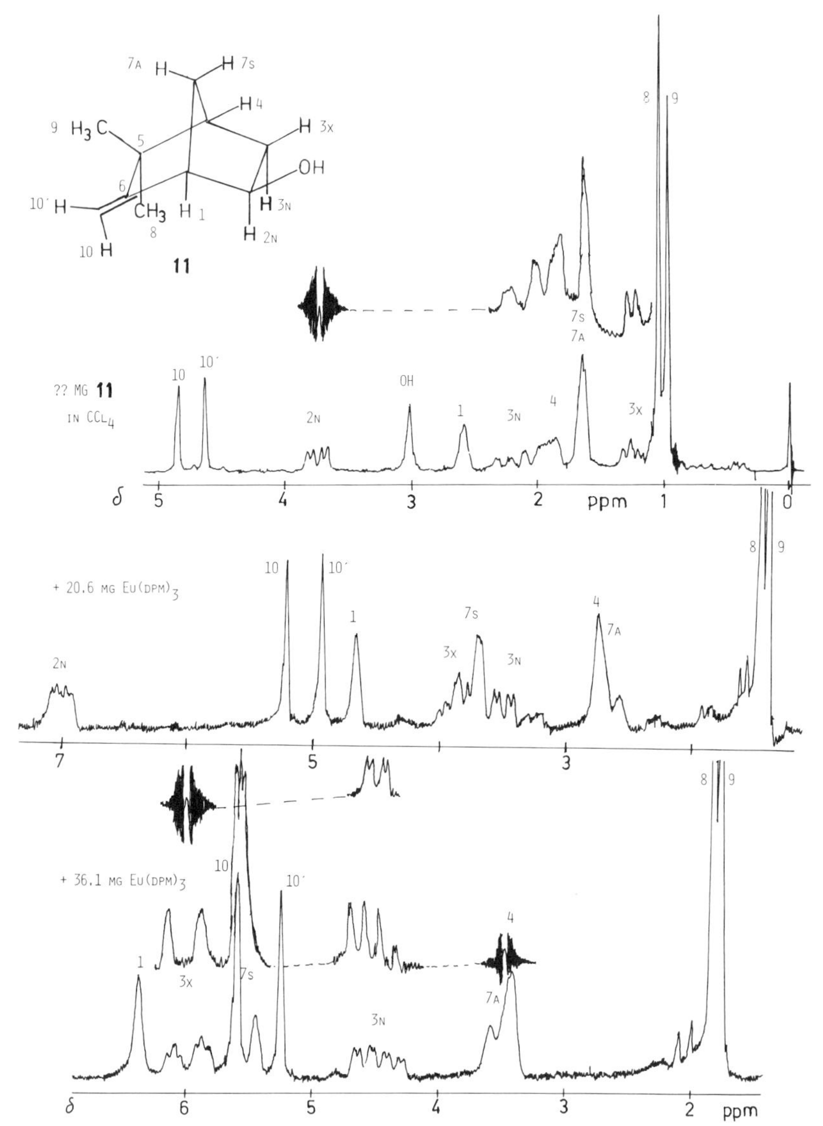

Figure 4-8. Proton NMR spectra of an unknown amount of Nojikigu alcohol, 5,5-dimethyl-6-methylenenorbornan-2-*exo*-ol (**11**), in 450 mg of CCl_4 (with TMS) before and after addition of 20.6 and 36.1 mg of $Eu(dpm)_3$; recorded on a 60-MHz Perkin Elmer Model R 12 B spectrometer at 33.5°C.

decoupling (see Figure 4-8). The dependence of the chemical shift of each proton on the added amount of $Eu(dpm)_3$ (Table 4-1) proved to be linear over a wide range. The shifts (in ppm) were plotted against added milligrams of $Eu(dpm)_3$ and the slopes of the lines were used as relative

TABLE 4-1. EVALUATION OF THE RELATIVE LIS VALUES (DEu) FOR NOJIKIGU ALCOHOL

Proton assign-ment	Eu(dpm)$_3$ added (mg)							Linear regression		
	0.0	7.5	14.5	20.6	26.7	36.1	48.6	Correlation coefficient	Slope	DEu
1	δ2.59	3.22	4.02	4.66	5.30	6.37	7.98	.99910	.105630	63.53
2*N*	3.74	4.81	6.00	7.01	8.03	9.72	12.30	.99938	.166265	100.00
3*N*	2.15	2.56	3.01	3.36	3.78	4.45	5.44	.99875	.063616	38.26
3*x*	1.15	1.97	3.01	3.80	4.63	5.97	8.02	.99929	.134662	80.99
4	1.86	2.11	2.45	2.74	3.02	3.41	4.10	.99882	.044030	26.48
7*s*	1.65	2.31	3.13	3.76	4.41	5.50	7.17	.99926	.107289	64.52
7*a*	1.65		2.34	2.66	2.97	3.52	4.27	.99889	.051532	30.99
8	1.04	1.16	1.32	1.44	1.57	1.78	2.13	.99887	.020694	12.45
9	0.97	1.10	1.26	1.38	1.51	1.73	2.08	.99903	.021116	12.70
10	4.84	4.95	5.11	5.23	5.34	5.57	5.88	.99760	.020275	12.19
10′	4.63	4.70	4.84	4.94	5.05	5.23	5.49	.99582	.017006	10.23
OH	3.02	7.15	11.98	15.94	19.92	26.42	36.32	.99946	.652501	392.45

[a] From ^{1}H NMR spectra in CCl_4–TMS with added amounts of Eu(dpm)$_3$ (see Figure 4-8). Amount of the substrate alcohol collected from GC was unknown. Chemical shifts (relative to TMS in ppm) are given for each proton in columns under total mg of LSR (added).

shift effect values, DEu. The value of α proton (2N) was used as the intramolecular standard of 100 relative units.

The structure 5,5-dimethyl-6-methylenenorbornan-2-*exo*-ol (**11**) for Nojikigu alcohol could be deduced from the ^{1}H NMR data based on experience with model compounds of the norbornan-2-*exo*-ol structure[9,28] and, further, by comparing the observed DEu values to those of α-isofenchol (**12**) and apoisofenchol (**13**).[15] Final verification was obtained by an iterative computer fitting of the pseudocontact eq. (4-1) to the experimental DEu values and atom coordinates. Willcox values[51] and measurement of Dreiding molecular models were used to evaluate the atom coordinates of **11** and **12**, assuming free rotation of the methyl group protons around the C—C axis and a C–O distance in the norbornanols of 1.46 Å. Parameters handled in computation of the average structure of a 1:1 complex are indicated in Figure 4-9. A computational procedure developed in our laboratory by Talvitie,[14] Häkli,[7] and Laatikainen,[52] titled LANTA (Appendix 1), operates as follows: first an estimate of the coordinates x, y, and z is made for the lanthanide atom and for each proton placed in the same coordinate system. The LIS (DEu) values are calculated by eq. (4-1). In order to maintain the constant term b near zero, a dummy proton far away (LIS = 0) can be included in computation. Residuals of the DEu values, equal to the differences between the observed and estimated ones, are minimized by a

11 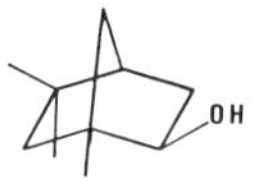 12  13

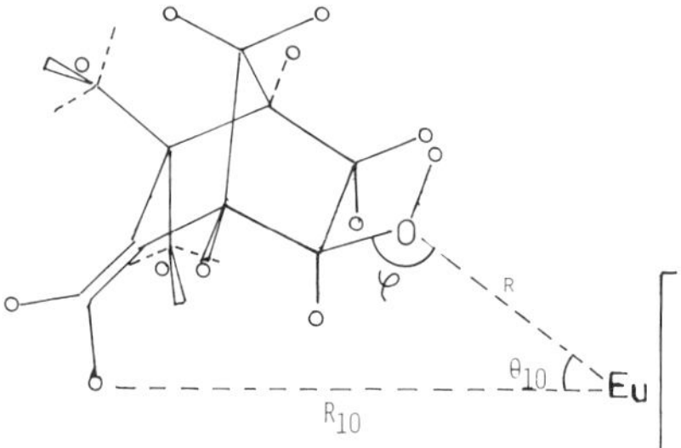

$$DEu_{EST.} = K \times \frac{3\cos^2\theta_I - 1}{R_I^3} + B \quad (1)$$

AVERAGE STRUCTURES OF THE 1:1 COMPLEXES

COMPOUND	φ^0	R Å	K	CORR.COEFF.
11	132.8	2.42	5066	0.99999
12	135.0°	2.19	4464	0.980 ×
12	129.7°	3.12	6480	0.999

× SECOND MINIMUM

	COMPOUND 11							COMPOUND 12			COMPOUND 13
I	$DEu_{OBS.}$	$DEu_{EST.}$	R_I Å	θ_I^0	X_I	Y_I	Z_I	$DEu_{OBS.}$	R_I Å	θ_I^0	$DEu_{OBS.}$
1	63.5	63.5	4.78	24.6	2.138	0.0	0.532				63.4
2N	100.0	100.1	3.64	35.3	1.152	1.123	-1.931	100.0	3.60	40.1	100.0
3X	81.0	80.8	4.23	29.5	-1.152	2.184	-0.472	81.5	4.47	31.6	75.4
3N	38.3	38.3	4.89	35.1	-1.152	1.123	-1.931	39.6	5.09	37.1	40.5
4	26.5	26.6	6.54	16.7	-2.138	0.0	0.532	22.9	6.96	24.1	25.4
7S	64.5	64.7	5.12	12.8	0.0	0.902	1.699	56.8	5.83	12.8	59.3
7A	31.0	30.8	6.51	6.6	0.0	-0.902	1.699	29.5	7.26	3.9	28.9
8	12.7	12.7	7.02	31.0	-1.250	-1.450	-2.300	11.9	7.28	35.9	12.4
9	12.5	12.5	7.94	16.3	-1.400	-2.500	-0.425	13.2	8.77	18.9	13.9
10	12.2	12.2	6.40	37.3	2.475	-1.925	-1.300	(FOR CASE R = 3.12 Å)			
10′	10.2	10.2	7.14	33.1	1.125	-2.525	-1.875				

Figure 4-9. Structural formulas of Nojikigu alcohol (**11**), α-isofenchol (**12**), and apoisofenchol (**13**). Computed average structure of the 1:1 complex between **11** and $Eu(dpm)_3$ in CCl_4 solution and comparison with the corresponding computation results for **12** and DEu values of **13**. For notation of the protons see Figure 4-8.

Newton-Raphson iteration method.[53] A group of linear equations is written for that purpose, in matrix notation:

$$\mathbf{D}\mathbf{\Delta} = \mathbf{C} \tag{4-2}$$

where **D** is the matrix of the partial differentials (Jacobian of the system), **Δ** is vector of the coordinates to be corrected, and **C** is the vector of the differences DEu_{obs}—DEu_{est}. The latter serve as corrections from which new coordinates are calculated by solving eq. (4-2) by a standard procedure:[53]

$$\mathbf{\Delta} = (\mathbf{D}^T\mathbf{D})^{-1}\mathbf{D}^T\mathbf{C} \tag{4-3}$$

The procedure is repeated with new coordinates until the residual sum of squares of the differences does not change. The system is strongly convergent and takes very little computer time.

Normally coordinates of the lanthanide atom, those of the heteroatom (oxygen in the case of **11**), and some protons are iterated and some others, preferably those whose stereoposition is known, are kept constant. Wrong minima may be met, but in the case of **11** (Figure 4-9) the result obviously is correct and its accuracy satisfactory. In contrast, model compound **12** gave two minima from which that one having shorter lanthanide–oxygen distance 2.19 Å apparently is unrealistic, but the other one, having an Eu–O distance of 3.12 Å, is realistic.

The computation for the 1:1 complex of $Eu(dpm)_3$ and **11** gave good results (Figure 4-9) by fixing the coordinates of protons 1, 2*N*, 3*x*, 3*N*, 4, 7*s* and 7*a* and iterating those for O and Eu in every round; the coordinates for protons 8, 9, 10, and 10′ were added to iteration one at the time.

Results of the LIS–NMR study for Nojikigu alcohol[16] demonstrate also that comparison of relative LIS values (DEu_{obs}; Figure 4-9) for unknown compound **11** and those of the structurally similar model compounds **12** and **13** in a "semiquantitative approach" can give very significant support of structure without computer fitting to the LSR–substrate complex.

The stereochemistry of a natural sesquiterpenoid 7-valeranol **14** was distinguished from an isomeric structure **15** by LIS–NMR using terpinen-4-ol **16** as a model compound (Figure 4-10). The substrate was isolated from *Valeriana officinalis L.* and shown to have the trans-decalin skeleton by elimination of water.[54] Proton NMR resolved the methyl group signals and olefinic proton peak, and addition of $Eu(dpm)_3$ led to more resolved spectra (examples in Figure 4-10), where some skeleton proton signals also could be assigned. The influence of the angle term of eq. (4-1) is very remarkable in this case because one of the methyl groups (13) and the other olefinic proton (12′) signals move toward high field when LSR is added, while the other signals move normally to low field.

The LIS effects of isopropenyl group protons (12, 12′, and 13) do not differ significantly for either structures **14** or **15** and so do not help in structure assignment. The effects of the other two methyl groups (14 and 15), however, should be very different, but there is ambiguity in their assignment. Consequently, LIS effects on the skeleton protons and a quantitative approach also must be used to make a firm decision between **14** and **15**.

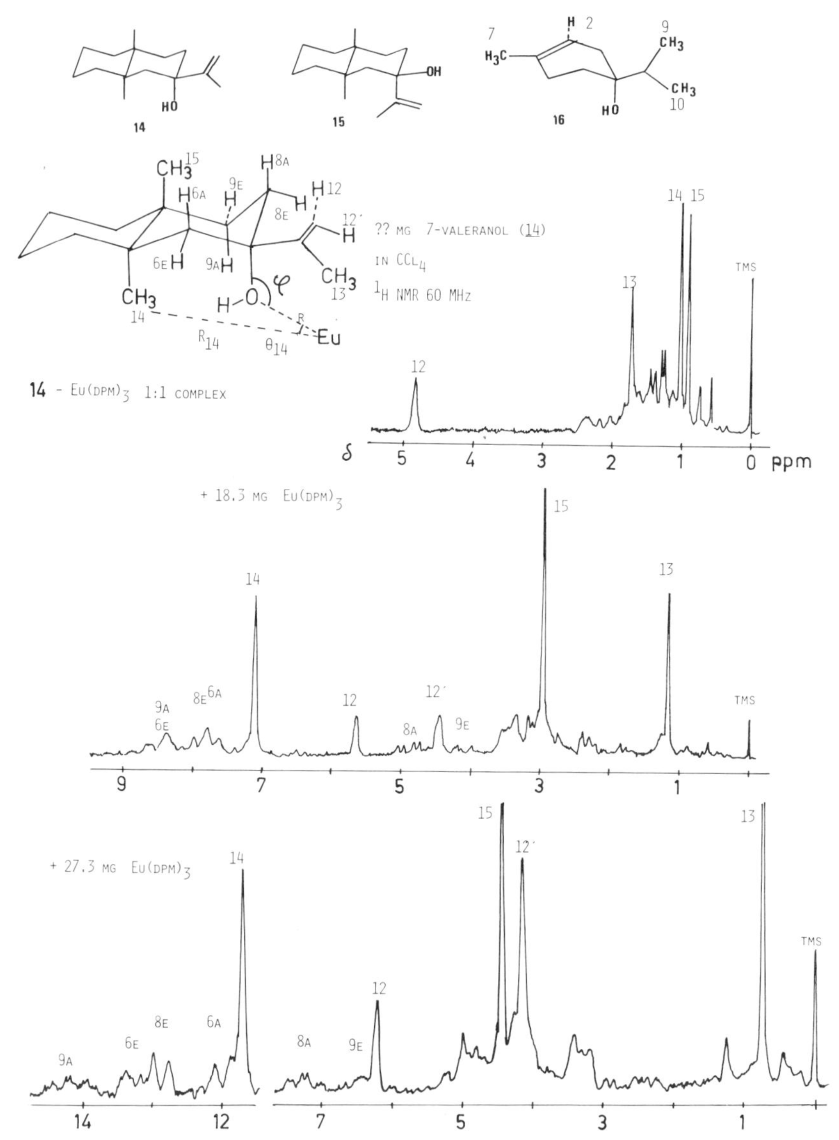

Figure 4-10. Structural formulas of the two possible structures **14** and **15** and conformation of terpinen-4-ol (**16**). Structural drawing of the 1:1 complex between **14** and $Eu(dpm)_3$. [1]Proton NMR spectra of an unknown amount of 7-valeranol in 450 mg of carbon tetrachloride (with TMS) at 33.5°C before and after additions of 18.3 and 27.3 mg of $Eu(dpm)_3$; recorded on a 60-MHz Perkin Elmer Model R 12 B spectrometer.

Computation of the fit of eq. (4-1) and 1:1 LSR–S complex for **14** and **15** gave results listed in Table 4-2.

The results favor structure **14**, but correlation coefficients are only 0.985 for **14** and 0.956 for **15**. The accuracy is not as good as for the more

TABLE 4-2. RESULTS OF THE COMPUTER ITERATION (LANTA) OF COMPOUNDS **14** AND **15**[a] (FIGURE 4-10)

	Compound **14**				Compound **15**			
i	DEu_{obs}	DEu_{est}	R_i	θ_i°	DEu_{obs}	DEu_{est}	R_i	θ_i°
6a	79.2	67.2	5.7	16	94.0	90.7	4.3	19.9
6e	94.0	94.6	4.9	22	79.2	79.4	3.6	40.0
8a	49.7	62.8	5.2	30	100.0	103.7	4.4	14.9
8e	91.4	88.0	3.8	42	91.4	90.7	3.8	34.9
9a	100.0	96.1	4.4	31	49.7	45.9	5.3	35.5
9e	42.8	41.1	5.7	34	42.8	44.1	5.9	28.5
12	12.4	20.7	5.2	48	12.4	19.1	3.9	54.9
12′	−4.5	−7.9	5.1	59	−4.5	−9.2	4.4	69.1
13	−6.3	−14.6	3.5	57	−6.3	−7.7	3.9	63.7
14	82.2	83.2	4.8	29	82.2	50.0	6.3	4.0
15	28.1	35.4	7.2	16	28.1	32.8	5.6	43.3

[a] Assumed to be, on the average, 1:1 complexes between the alcohol and $Eu(dpm)_3$.

rigid norbornanols. This arises partly by possible skewing from the chair conformation in the more substituted skeleton ring caused by steric repulsion of the substituents. To improve the reliability of the structure deduction, comparison with model compounds were made. Computation for terpinen-4-ol **16** fits very well (correlation coefficient 0.999) with an *endo*-OH half-chair conformation (drawn in Figure 4-10) supporting structure **14** for 7-valeranol. More support for these structural conclusions is obtained from computations for tertiary norbornanols and norbornenols (Table 4-3).

TABLE 4-3. PARAMETERS FROM COMPUTATIONS FOR THE AVERAGE STRUCTURES OF 1:1 COMPLEXES BETWEEN EU(DPM)$_3$ AND SOME TERTIARY ALCOHOLS[a]

Compound/structure	R	φ°	Correlation coefficient
Structure **14** for 7-valeranol	2.6	123	0.985
Structure **15** for 7-valeranol	1.7	142	0.956
Terpinen-4-ol (**16**)	3.1	152	0.999
3-Methyl-3-nortricyclanol	2.8	158	0.983
2-*endo*-methylnorbornan-2-*exo*-ol (**5**)	1.9	152	0.997
2-*endo*-5,5-trimethylnorbornan-2-*exo*-ol	2.1	149	0.996
2-*endo*-methylisoborneol (MIB)	3.2	155	0.998
2-*exo*-methylnorbornan-2-*endo*-ol (**6**)	1.8	165	0.993
2-*exo*-5,5-trimethylnorbornan-2-*endo*-ol	2.8	170	1.000
2-*endo*-methyl-5-norbornen-2-*exo*-ol (**3**)	1.8	154	0.992
1,2-*endo*-7,7-tetramethyl-5-norbornen-2-*exo*-ol	3.3	152	0.998
2-*exo*-methyl-5-norbornen-2-*endo*-ol (**4**)	2.3	157	0.994

[a] In CCl_4 solutions from observed DEu values.[7,15,30,37,54] For definitions of R and φ see Figure 4-9 and 4-10.

Comparison of the structural parameters for average 1:1 complexes (Table 4-3) shows that structure **15** for 7-valeranol leads to a minimum where the Eu–O distance is unrealistically short (1.7 Å), whereas in *exo*-OH norbornyl structures this distance is between 1.8 and 3.3 Å, depending on the amount of alkyl substitution (crowding). Moreover, the correlation for **15** is significantly lower (0.956) than for model compounds (0.983–1.000). In contrast, the result for structure **14** gives rise to a distance of 2.6 Å and a correlation coefficient of 0.985, both within the range for model tertiary alcohols (Table 4-3).

A quantitative approach alone cannot distinguish between possible stereoisomers in all cases, as is obvious in the case of 7-valeranol above. One reason is ambiguity in proton signal assignments. Lanthanide-induced shift effects, however, can aid in signal assignment. For example Willcott et al[55] use an autoassignment program together with computation of complex structures from low-precision LIS-effect data and are able to distinguish between four stereoisomeric methylbicyclo[2.2.2]octenols at a 98% or greater confidence level.

Primary Alcohols: Diastereotopic Groups

In the quantitative LIS–NMR study of cyclic methanols (−)-myrtenol (**17**), 3-cyclopentenyl-1-methanol (**18**), and 5-norbornene-2-*endo*-methanol (**19**), we observed[17] considerable contact effects for the α protons in CH_2 groups of europium complexes and marked differentiation of these protons caused by their diastereotopy (Figure 4-11). Only a few months later, Abraham et al. reported[56] similar behavior for two relatives of **19**, 3-*exo*-methyl-5-norbornene-2-*endo*-methanol (**20**), and 3-*endo*-methyl-5-norbornene-2-*exo*-methanol (**21**).

In contrast with the secondary and tertiary polycyclic alcohols, where OH is located at a ring carbon (see above), compounds **17**–**21** have additional structural complexity because of possibility of several stable (staggered) conformations involving the carbon atom of the CH_2OH group. If the rest of the molecule is asymmetric, the protons of this methylene group are diastereotopic and may resolve in NMR. In addition the equilibrium population of the conformations may be different. Also the effect of the shift reagent can be different for these two protons. This was actually observed[17] in the case of **19** but not for **17**, **20**, or **21**.[56]

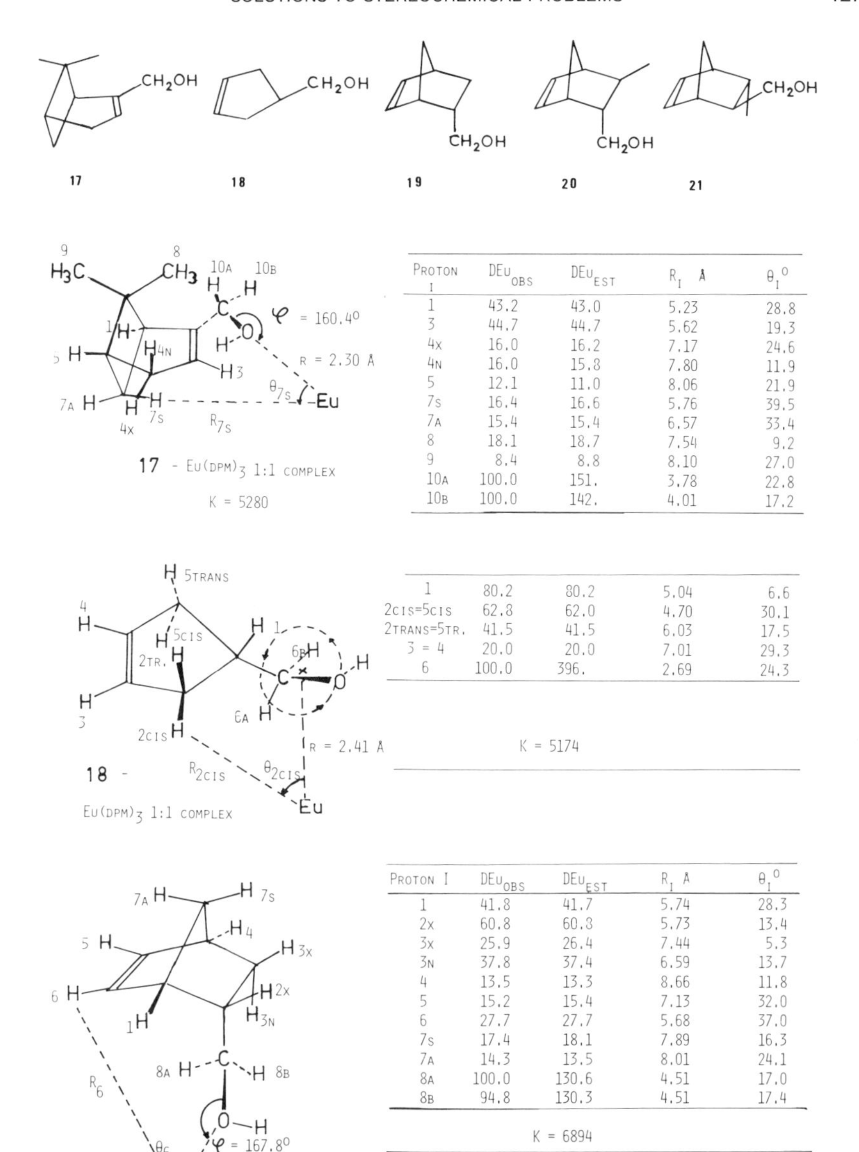

Proton I	DEu_{OBS}	DEu_{EST}	R_I Å	θ_I °
1	43.2	43.0	5.23	28.8
3	44.7	44.7	5.62	19.3
4x	16.0	16.2	7.17	24.6
4N	16.0	15.8	7.80	11.9
5	12.1	11.0	8.06	21.9
7s	16.4	16.6	5.76	39.5
7A	15.4	15.4	6.57	33.4
8	18.1	18.7	7.54	9.2
9	8.4	8.8	8.10	27.0
10A	100.0	151.	3.78	22.8
10B	100.0	142.	4.01	17.2

1	80.2	80.2	5.04	6.6
2cis=5cis	62.8	62.0	4.70	30.1
2trans=5tr.	41.5	41.5	6.03	17.5
3 = 4	20.0	20.0	7.01	29.3
6	100.0	396.	2.69	24.3

K = 5174

Proton I	DEu_{OBS}	DEu_{EST}	R_I Å	θ_I °
1	41.8	41.7	5.74	28.3
2x	60.8	60.3	5.73	13.4
3x	25.9	26.4	7.44	5.3
3N	37.8	37.4	6.59	13.7
4	13.5	13.3	8.66	11.8
5	15.2	15.4	7.13	32.0
6	27.7	27.7	5.68	37.0
7s	17.4	18.1	7.89	16.3
7A	14.3	13.5	8.01	24.1
8A	100.0	130.6	4.51	17.0
8B	94.8	130.3	4.51	17.4

K = 6894

Figure 4-11. Structural formulas of five cyclic methanols **17**–**21** (names in the text) and computed average structures of the dynamic 1:1 complexes of **17**–**19** and $Eu(dpm)_3$. Three staggered conformations of 5-norbornene-2-*endo*-methanol (**19**) are illustrated in the bottom right of the figure.

Results of computations for a 1:1 complex of LSR–substrate are presented in Figure 4-11. Our normal procedure, using the LANTA program, gave satisfactory results for **17** and **19**, which indicates there is some less populated rotamer (conformer) in the equilibrium. In the case of **19** it clearly is **19b**, which is also seen in the different DEu values of the well-separated CH_2 protons. The less crowded and more symmetric molecule **18** in the LANTA iteration gave a perfect fit with an average structure of a rapidly rotating CH_2OH group, the C—C bond being the rotation axis. Thus the effective location of the oxygen atom (x, Figure 4-11, middle drawing) is in the center of the rotation circle of the O. Populations of the staggered conformers must be about equal and in rapid equilibrium at this temperature (33.5°C).

Abraham and co-workers[56] used a modified LSR-substrate complex in computing for NMR data of **20** and **21**, where populations of the possible conformers were weighted. The best fit with pseudocontact eq. (4-1) was obtained by **20a**:**20b**:**20c** = 67:0:33 (structures a, b, and c as drawn for **19** in Figure 4-11) and by **21a**:**21b**:**21c** = 33:0:67, the reverse of that for **20**.

The fit with the pseudocontact effect for computing structures **17**–**21** was perfect when α protons of the CH_2 group were omitted. The estimated pseudocontact LIS effect values (DEu_{est}, Figure 4-11) for these protons also were calculated using eq. (4-1). They were all considerably larger than the observed values (DEu_{obs}). The difference corresponds to a negative contact effect of europium on α protons. This is consistent with the predictions for europium:[3,4,57] the contact shift of the Eu complex should be negative but the dipolar shift positive for θ angles below 54.74° (the magic angle). In the present case we calculate the contact shifts (in DEu values) to be 51* and 42 relative units on the α protons of **17**, 296 relative units for **18**, and 30.6 and 30.3 relative units for **19**.

Diastereotopy has an important role in the LIS–NMR of aliphatic chain structures. Proton NMR spectra of aliphatic acyclic alcohols can be simplified dramatically and resolved by use of LSRs; differentiation of the diastereotopic geminal protons can be greatly improved.[58] Generally, differential LSR shifting of the α protons is small.[59] In contrast, other aliphatic diastereotopic protons usually have larger differences in their structural environments in each rotamer and so exhibit large differential shifts. For example, the γ protons in 2,3,3-trimethyl-1-pentanol and the δ protons in 3,5,5-trimethyl-1-hexanol and 2,2,3-trimethyl-1-

* See Figure 4-3: 151 − 100 = 51.

pentanol showed a marked nonequivalency and large differences in LIS effects.[58] On the other hand, the γ protons of the formally similar 2-methyl-2-propyl-1-pentanol show practically no differential shift. All these phenomena can be rationalized readily by examining the possible conformers and by (semi)quantitative treatment of the NMR data.[58,60]

Ring Conformations

Determination of stereochemistry, especially ring conformations, is central in LIS–NMR studies of natural products.[61] One of the most common rings, cyclohexane, has been scrutinized by studies of numerous model compounds, such as cyclohexanols and cyclohexanones.[18] The chair conformation of cyclohexane can be investigated for its properties in compounds where it is fixed, eg, in adamantane derivatives.[61–63] Flexible cyclohexanes lead to more complexity in LIS–NMR spectra, whereas several stable conformers in may equilibrate. In NMR solvents, the cyclohexanol (**22**, Figure 4-12) system contains both equatorial (**22a**) and axial (**22b**) forms.[64] To determine the conformer ratio, LIS–NMR data for compounds having essentially only one dominating conformer, such as in *trans*- (**23**) and *cis*- (**24**) 4-*t*-butylcyclohexanol,[11] could be used.[64] The observed shift effect can be related to the equatorial–axial (eq–ax) equilibrium according to the expression:

$$DEu_{obs} = N_{eq} \cdot DEu_{eq} + N_{ax} \cdot DEu_{ax} \qquad (4\text{-}4)$$

Assignments for cyclohexanol (**22**) could be verified[64] by comparing the spectra of *trans*-3-*trans*-5-dideuterocyclohexanol.[65] Effects of $Eu(dpm)_3$ on proton signals of cyclohexanols **22**, **23**, and **24** calculated for a molar LSR–substrate ratio of 1:1 (ΔEu, ppm) in $CDCl_3$ are presented in Table 4-4.

The conformer ratio $N_{eq}:N_{ax} = 73:27$ yields an equilibrium constant ($K_{27.4} = 2.70$) for cyclohexanol in the absence of hydrogen bonding effects, because it has been observed that metal–alcohol complexation has a negligible effect upon the eq–ax ratio.[64] Other methods for determining this equilibrium suffer from drawbacks; for example, dynamic NMR (lineshape) methods necessitate the use of low temperatures, whereas hydrogen bonding alters the eq–ax ratio.

4-*t*-Butylcyclohexanols, **23** and **24**, and their methyl derivatives **25–30** (Figure 4-12) were useful models for stereochemical studies involving

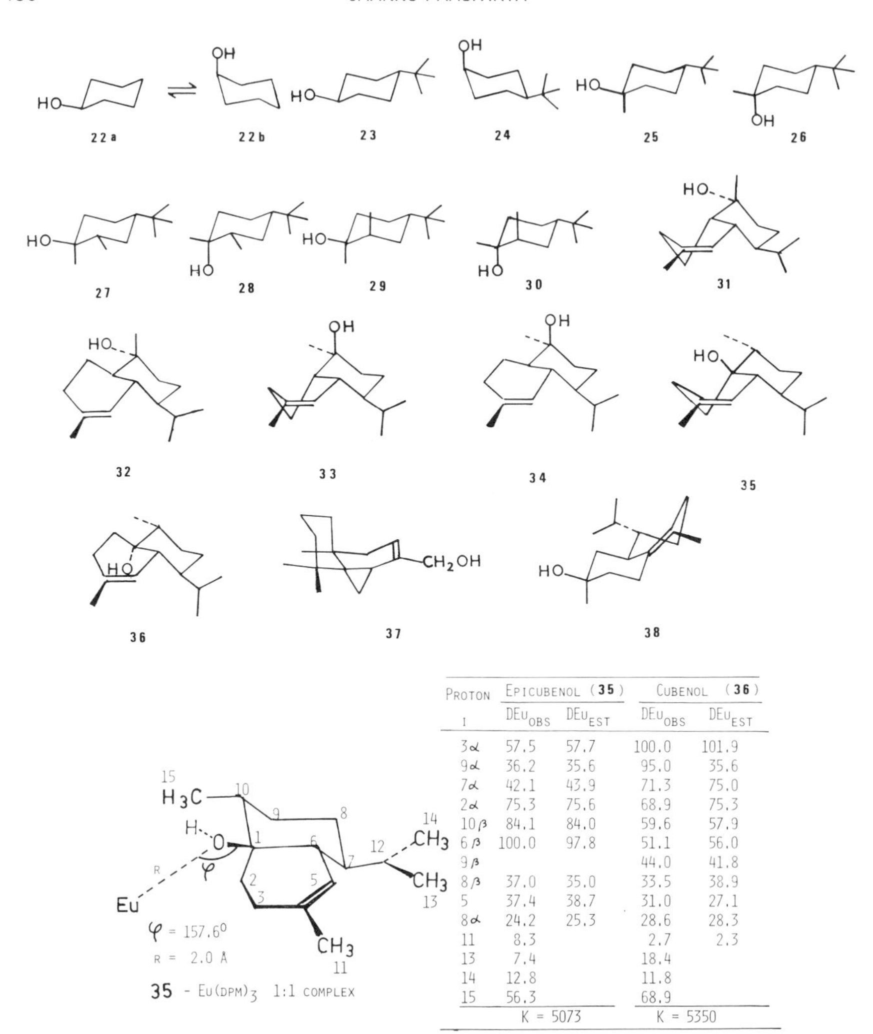

Proton I	Epicubenol (35) DEu$_{OBS}$	Epicubenol (35) DEu$_{EST}$	Cubenol (36) DEu$_{OBS}$	Cubenol (36) DEu$_{EST}$
3α	57.5	57.7	100.0	101.9
9α	36.2	35.6	95.0	35.6
7α	42.1	43.9	71.3	75.0
2α	75.3	75.6	68.9	75.3
10β	84.1	84.0	59.6	57.9
6β	100.0	97.8	51.1	56.0
9β			44.0	41.8
8β	37.0	35.0	33.5	38.9
5	37.4	38.7	31.0	27.1
8α	24.2	25.3	28.6	28.3
11	8.3		2.7	2.3
13	7.4		18.4	
14	12.8		11.8	
15	56.3		68.9	
	K = 5073		K = 5350	

Figure 4-12. Ring conformations of cyclohexanol model compounds **22**–**30** and natural sesquiterpene alcohols **31**–**38** (names in the text). Computed average structure of the dynamic 1:1 complex of epicubenol (**35**) and Eu(dpm)$_3$ and comparison of the LIS values of epicubenol (**35**) and cubenol (**36**).

TABLE 4-4. ΔEu Values (ppm) for Cyclohexanols in $CDCl_3$.[64]

Compound	H_1	H_2 cis	H_2 trans	H_3 cis	H_3 trans
trans-4-*t*-Butylcyclohexanol (**23**)	21.7	14.7	13.6	4.4	5.4
cis-4-*t*-Butylcyclohexanol (**24**)	24.7	14.9	8.2	13.6	6.7
Cyclohexanol (**22a** ⇌ **22b**) observed	22.5	16.1	13.1	6.9	5.8
Cyclohexanol calculated for **a**:**b** 73:27	22.5	14.8	12.1	6.9	5.8

ring conformation determinations of a series of natural sesquiterpenoid alcohols **31**–**38** (Figure 4-12) by LIS–NMR.[14,66–69] The compounds studied were torreyol (**31**), α-cadinol (**32**), *T*-muurolol (**33**), *T*-cadinol (**34**), epicubenol (**35**), cubenol (**36**), the substituted methanol product of the reduction of hinokiic acid (**37**) and 4α-hydroxy-10(1)-cadinene (**38**). The structure deductions were done by semiquantitative comparison of the LIS effects and coupling patterns of the proton NMR signals of models and studied compounds and then verified by determining the average structures of the dynamic shift reagent–substrate complexes by iterative computing (LANTA program). A typical computing result[14,67] is illustrated in the lower part of Figure 4-12.

The same treatment was successful for cyclic ketones. For example, a perfect fit of the compound LSR-substrate 1:1 dynamic average complex structure with the pseudocontact equation was obtained for valeranone (**39**, Figure 4-13),[70] after 4-*t*-butylcyclohexanone (**40**) and 2-*cis*-methyl-4-*t*-butylcyclohexanone (**41**) had been used as models. Actually the computed complex structures of ketones represent averages of complexation that occurs from two directions on the different lone electron pairs of the oxygen atom.[71] The computed structure can be considered a weighted average in favor of the sterically less hindered[72] direction.

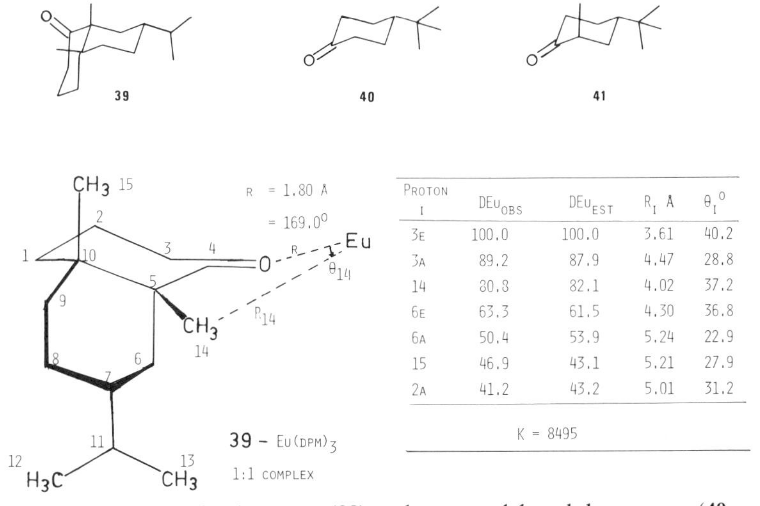

PROTON I	DEu_{OBS}	DEu_{EST}	R_I Å	θ_I °
3E	100.0	100.0	3.61	40.2
3A	89.2	87.9	4.47	28.8
14	80.8	82.1	4.02	37.2
6E	63.3	61.5	4.30	36.8
6A	50.4	53.9	5.24	22.9
15	46.9	43.1	5.21	27.9
2A	41.2	43.2	5.01	31.2

K = 8495

Figure 4-13. Structures of valeranone (**39**) and two model cyclohexanones (**40** and **41**). Computed average structure of the dynamic 1:1 complex of valeranone (**39**) and Eu(dpm)$_3$. Actual coordination of Eu has been possible from two sites of the oxygen atom; the structure obtained by computation represents a statistical average of the complexes with Eu–O bonds at both sites and is in favor of the nonmethylated site.

Stereochemistry of the Deuterium Label

The course of the steric addition reactions to norbornenes and nortricyclenes has been of great interest in, for example, exploring the nonclassical ion problem.[21] If the reactant contains active hydrogen this can be replaced by deuterium. Then the location of deuteron in the product gives additional information on the steric course of the reaction. In many analyses of the location of such isotope labels LIS–NMR is a significant tool.

The catalytic deuteration of the double bond of norbornenes has been found to be an exo-exo-cis addition.[73] Analysis of the product from dehydronorcamphor by LIS–NMR is illustrated in Figure 4-14. The $Eu(dpm)_3$-shifted ^1H NMR spectra (see upper part of Figure 4-14) were readily assigned and comparison to the nondeuterated norcamphor showed that signals 5*x* and 6*x* were missing. Computations for the average 1:1 LSR–substrate dynamic complex (lower part of Figure 4-14) gave a very good correlation with the structure norcamphor-5-*exo*-

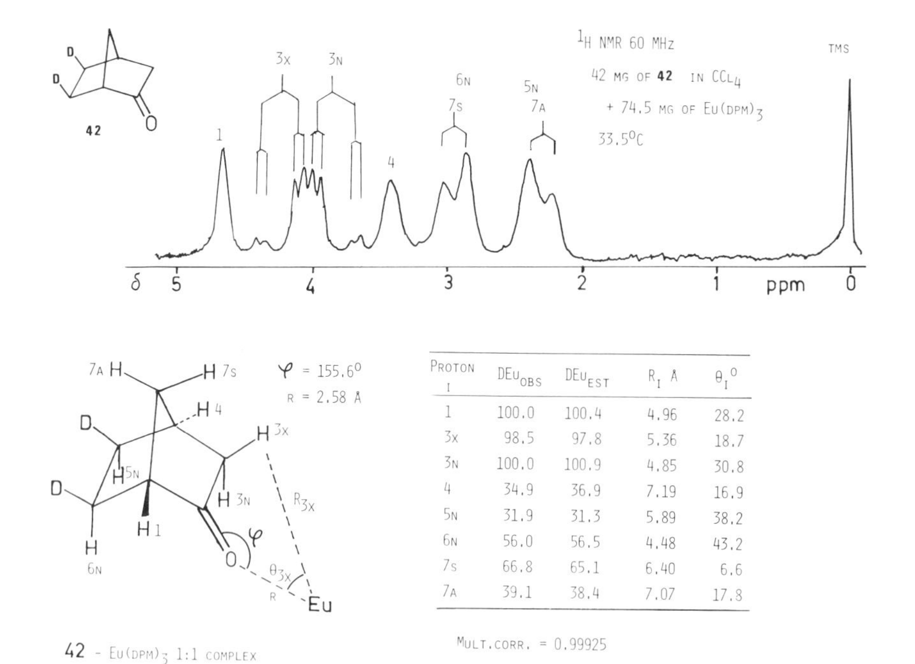

PROTON I	DEu$_{OBS}$	DEu$_{EST}$	R$_I$ Å	θ$_I$°
1	100.0	100.4	4.96	28.2
3x	98.5	97.8	5.36	18.7
3N	100.0	100.9	4.85	30.8
4	34.9	36.9	7.19	16.9
5N	31.9	31.3	5.89	38.2
6N	56.0	56.5	4.48	43.2
7s	66.8	65.1	6.40	6.6
7A	39.1	38.4	7.07	17.8

MULT.CORR. = 0.99925

Figure 4-14. Proton NMR spectrum (60 MHz) and computed average 1:1 complex structure of the product of catalytic deuteration of dehydronorcamphor. Signal assignments can be verified by spin decoupling. For example, when proton 4 is irradiated the smaller coupling of 3*x* vanishes, and when 7*a* is irradiated the smaller coupling of 3*n* vanishes. Spectrometer: Perkin Elmer R 12 B.

6-*exo*-d_2 (**42**). Another computation for an alternative structure for the product as norcamphor-5-*endo*-6-*endo*-d_2, resulted in clear disagreement with observed shift data.[73]

Addition of acids to norbornene (**43**) and nortricyclene (**44**) yield 2-*exo*-norbornyl derivatives the stereochemistry of which was accepted originally as support for Winstein's[74] nonclassical 2-norbornyl cation (**45**). This cation had been proposed as an intermediate structure in both addition and solvolysis reactions in solution and also to explain the spectral properties of the 2-norbornyl ion in superacid media.[75–77] An alternative structure for this intermediate is a classical 2-norbornyl carbenium ion (**46a**), which is energetically identical to and in rapid equilibrium with its Wagner–Meerwein rearrangement product. In the latter case, an explanation for the dominance of the exo product is obtained from steric conditions of **46**. This is the famous suggestion of H. C. Brown.[21,78–80]

Addition of formic acid to norbornene (**43**) and nortricyclene (**44**) yields norbornane-2-*exo*-formate as the only product.[81–84] The steric course of these reactions could be studied by using deuterated acid and LIS–NMR of the products (Figure 4-15).[85] The formate products were saponified by KOH–methanol (where no rearrangements take place) and the resulting deuterated norbornan-2-*exo*-ols (**1a**–**1d**) could be scrutinized for stereolocation of the deuterium label by comparing their $Eu(dpm)_3$-shifted proton NMR spectra with those of the nondeuterated norbornane-2-*exo*-ol (**1**, Figure 4-1). In the product from **43** signals 3*x* and 7*s* were smaller than the others; according to the integral, they were each 1/2-proton signals (see upper part of Figure 4-15). Clearly, the 2-norbornyl cation has been formed by addition of the deuteron to the double bond from the exo direction and Wagner–Meerwein rearrangement (only) has taken place before nucleophilic attack by the formate anion. Even distribution of D between 3*x* and 7*s* comes either from structure **45a** (a symmetrical nonclassical ion[74]) or from a rapid equilibrium of some asymmetric pair of ions (eg, **46a** and **46b**). Similarly, from nortricyclene (**44**, see middle part of Figure 4-15) the 2-norbornyl ion forms by addition of a deuteron (proton) to the corner (C6) of the cyclopropane ring and again the observed ratio of stereopositions 6*x* and 6*n* is 50:50. This could also be explained by one symmetrical ion (**45b**) or by a rapidly exchanging pair (eg, **46c** and **46d**) of asymmetric ions. Although the symmetric label distribution in our results, in the opinion of this author, is a clear indication of the existence of σ-bridging in 2-norbornyl ion, it cannot be taken as evidence of totally symmetric structure **45**. Also the even distribution is not measured very accurately by the NMR integral. For the reactions of norbornene with other acids,[86–91]

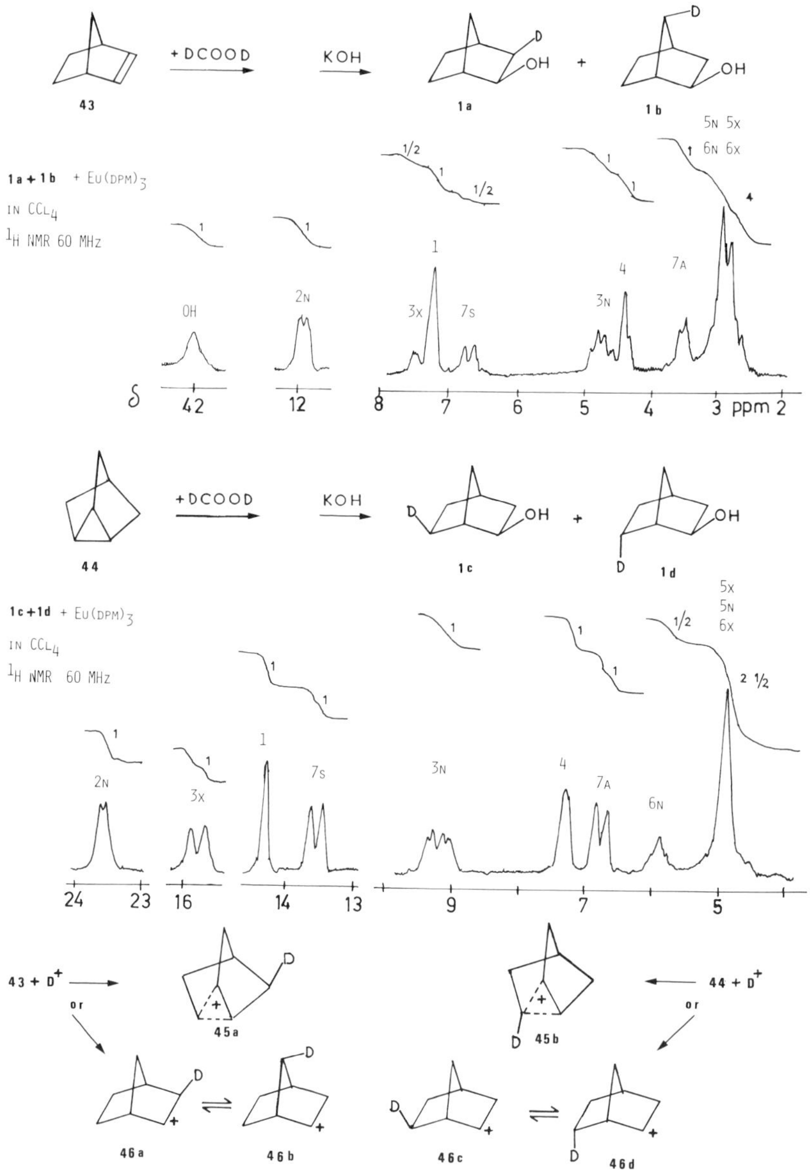

Figure 4-15. The Eu(dpm)$_3$-shifted spectra of saponified reaction products of norbornene (**43**) and nortricyclene (**44**). The location of deuterium is observed by comparison with the spectra of nondeuterated norbornan-2-*exo*-ol (**1**, Figure 4-1) and by integration of the signals. The bottom of the figure shows structures of the cations (incumbered in ion pairs) formed in the primary protonation (addition of D^+) according to the theory of Winstein (nonclassical ions **45**) or according to suggestion of Brown (equilibrating classical ions **46**), which by trapping by $DCOO^-$ lead to the ester precursors of **1a**, **1b**, **1c**, and **1d**. Spectrometer: Perkin Elmer R 12 B.

as well as in the reactions of nortricyclene with other acids,[91–93] uneven distribution of label was observed in most cases. The LIS–NMR method also was used to aid the determination of the deuterium label positions[89,90] as an alternative to high-field proton NMR.[87,91,93] In these studies more D was at 3*x* than at 7*s* from products of addition to norbornene (**43**) and H. C. Brown and co-workers claimed this as evidence of the capture of the 2-norbornyl cation in an unsymmetrical (classical **46**) state.[86–89,93] This statement cannot be accepted, because in all cases of uneven 3*x*–7*s* distribution, deuterium also has moved to other positions; this was very clearly seen in reactions at higher temperatures.[90] The same applies to the reaction of nortricyclene, where the uneven D distribution between 6*x* and 6*n* (in favor of 6*n*) can be associated with the stable edge-protonated nortricyclene ion[91] and to further rearrangement of deuterium[93] by other than Wagner–Meerwein rearrangements occuring at lengthened lifetimes of the norbornyl ion. According to Nordlander and co-workers,[94] deuteronation of norbornene (**43**) produces a bridged ion unsymmetrically associated with its gegenion and this also can explain an uneven distribution of deuterium in the product without invoking classical intermediate structures **46a**, **46b**. This explanation fits with recent results of the kinetic studies on solvolysis of different 2-norbornyl derivatives,[95–97] which have led to the conclusion that 2-norbornyl ion is actually σ bridged but asymmetric (**47** at upper left corner of Figure 4-16).

The "graded σ participation" of the norbornyl cation structures introduced by Grob and co-workers[95–97] is analogous to the "semi-nonclassical" structure **48** (Figure 4-16) of the tertiary 2-methylnorbornyl ion. This was first drawn by this author in 1965[98] to explain the fast reaction rate and the exo structure (**51**) of the only primary product of the reaction of 2-*endo*-methyl-norbornan-2-*exo*-ol (**5**) in formic acid. In 1969 this structure was specified by Olah and co-workers[99] from spectra taken in superacids. Brown and co-workers have shown marked similarity in the reactivities of 2-methylnorbornyl with 2-norbornyl compounds; they have explained this by the classical nature of the intermediates in both systems.[21] However, recent solvolytic studies show the "graded σ participation" in 2-methylnorbornyl compounds analogously to norbornyl reactions[100] and therefore support the original idea of intermediate **48** given by the present author.[98] Consequently, study of the steric course of the additions of acids to methyl and methylene norbornanes and nortricyclenes has been of great interest to us.[85,101] The LIS–NMR combination is a valuable tool in determining the location of the deuterium label in products. As a simple example, analysis of the products

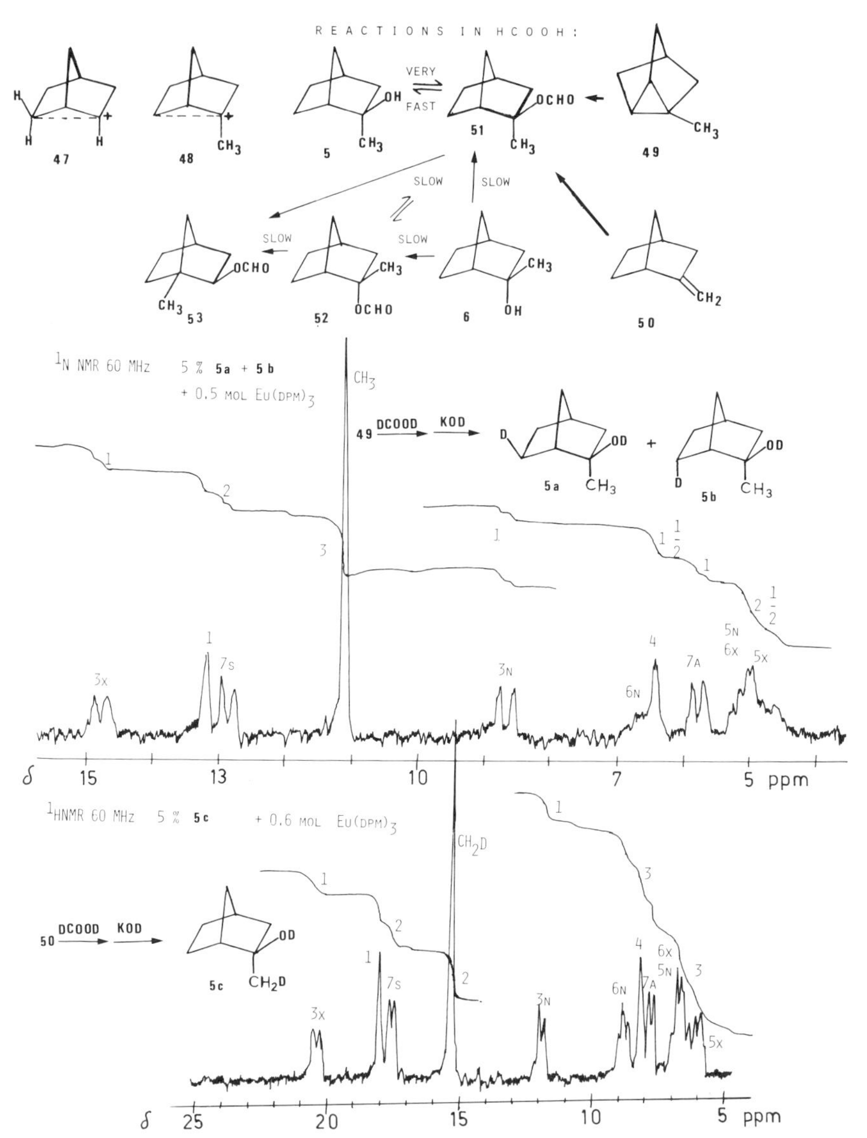

Figure 4-16. Graded σ-bond participation structures of 2-norbornyl (**47**) and 2-methylnorbornyl (**48**) cations. Scheme of the reactions of 2-*endo*-methylnorbornan-2-exo-ol (**5**), 2-exo-methylnorbornan-2-endo-ol (**6**), 1-methylnortricyclene (**49**), and 2-methylenenorbornane (**50**) in formic acid. The LIS–NMR spectrum (middle part) of the saponified reaction product of 1-methylnortricyclene (**49**) and dideuteroformic acid. The LIS–NMR spectrum (lower part) of the saponified reaction product of 2-methylenenorbornane (**50**) and dideuteroformic acid. Spectrometer: Perkin Elmer R 12 B, 33.5°C.

of the addition of deuterated formic acid to 1-methylnortricyclene (**49**) and to 2-methylenenorbornane (**50**) is illustrated in Figure 4-16. As expected, deuterated **51** was the only primary product found in both reactions, and the reaction kinetics differed very much.[101] Product analyses were done on the saponified primary products by Eu(dpm)$_3$-shifted proton NMR spectra. Assignments were performed by comparing the spectra to those of nondeuterated alcohol **5** (Figure 4-5). The only norbornane skeleton signals that were smaller than the others indicating partial deuteration were 6*n* and 6*x* in the product from **49** and methyl signal in the product from **50**. Corner protonation of **49**, similar to that in the case of nortricyclene (**44**), had taken place (see also Figure 4-15). This is support for a similar nature for both the 2-norbornyl (eg, **47**) and the 2-methylnorbornyl (eg, **48**) ions. Reaction of 2-methylenenorbornane (**50**) apparently started by addition of the deuteron to the olefinic CH_2 carbon, producing the 2-methylnorbornyl ion (methyl deuterated **48**), and primarily yielding tertiary *exo*-formate (deuterated **51**) by addition of formate anion without rearrangement. This interesting phenomenon can be used as a model for studies of the reactions of camphene and other bicyclic terpenes and their tricyclic relatives.[85]

Tricyclene (**54**) and camphene (**56**) react in formic acid to form one single product: isobornyl formate (**58**; see top of Figure 4-18).[81,85] Using dideuteroformic acid and LIS–NMR to detect the stereopositions of the deuterium label in the product (Figure 4-17) is a new way to get information about the famous cation from camphene.[85] This intermediate actually is the very first carbocation suggested by Meerwein and van Emster in 1922 for rearrangement of camphene hydrochloride to isobornyl chloride.[102] The cation has a structural relation to the tertiary 2-methylnorbornyl ion (see above) and, remarkably, it is also the very first ion for which a "nonclassical" structure has been suggested. Nevell and co-workers in 1939 specified it as a "mesomeric" cation, probably following a suggestion of C. K. Ingold.[103]

The Eu(dpm)$_3$-shifted ^{1}H NMR spectra of saponified products from the addition of dideuteroformic acid to **55** and **56** (examples in Figure 4-17) could be assigned by comparing them to LIS–NMR spectra of the nondeuterated isoborneol.[10] The peak heights and integrals show very clearly that methyl signals 9 and 10 are smaller than 8 both in product **55** from tricyclene (**54**) and in **57** from camphene (**56**). A more careful integration would show that signals 5*x*, 5*n*, 6*x*, and 6*n* are somewhat smaller than the other skeleton proton signals in **55** but not in **57**. Running deuterium NMR spectra of the products (bottom right in Figure 4-17) verifies the observations that deuterium has been incorporated

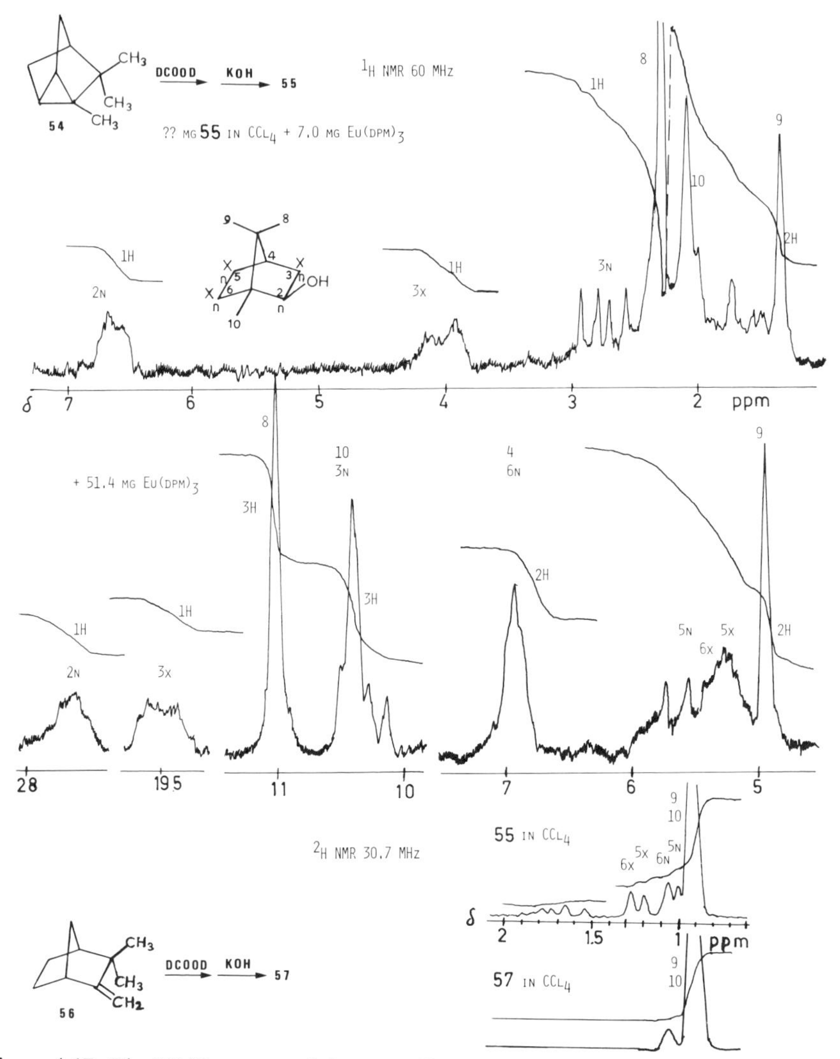

Figure 4-17. The NMR spectra of the saponified reaction products of tricyclene (**54**) and camphene (**56**) with dideuteroformic acid. Upper and middle spectra are Eu(dpm)$_3$-shifted spectra of deuterated isoborneol (**55**) from reaction of **54** (recorded on a Perkin Elmer Model R 12 B proton NMR spectrometer at 33.5°). The lower spectra are deuterium NMR spectra of **55** and of otherwise deuterated isoborneol (**57**) from reaction of **56**, respectively, recorded on a Varian XL 200 spectrometer.

in the methyl groups in both products and to several skeleton positions of product **55** from tricyclene (**54**). In ^{13}C NMR spectra deuteration was observed at carbons 9 and 10 of both **55** and **57** but in carbons 5 and 6 of **55** only. Multiple deuteration of the products was also verified by mass spectrometry.[85]

These results can be explained by rapid stereospecific rearrangements of tertiary cations in which graded σ participation (see above) may be one factor causing the specificity (Figure 4-18) in the form of exo attack of nucleophile (anion or nonbonded electrons) to the cationic center.

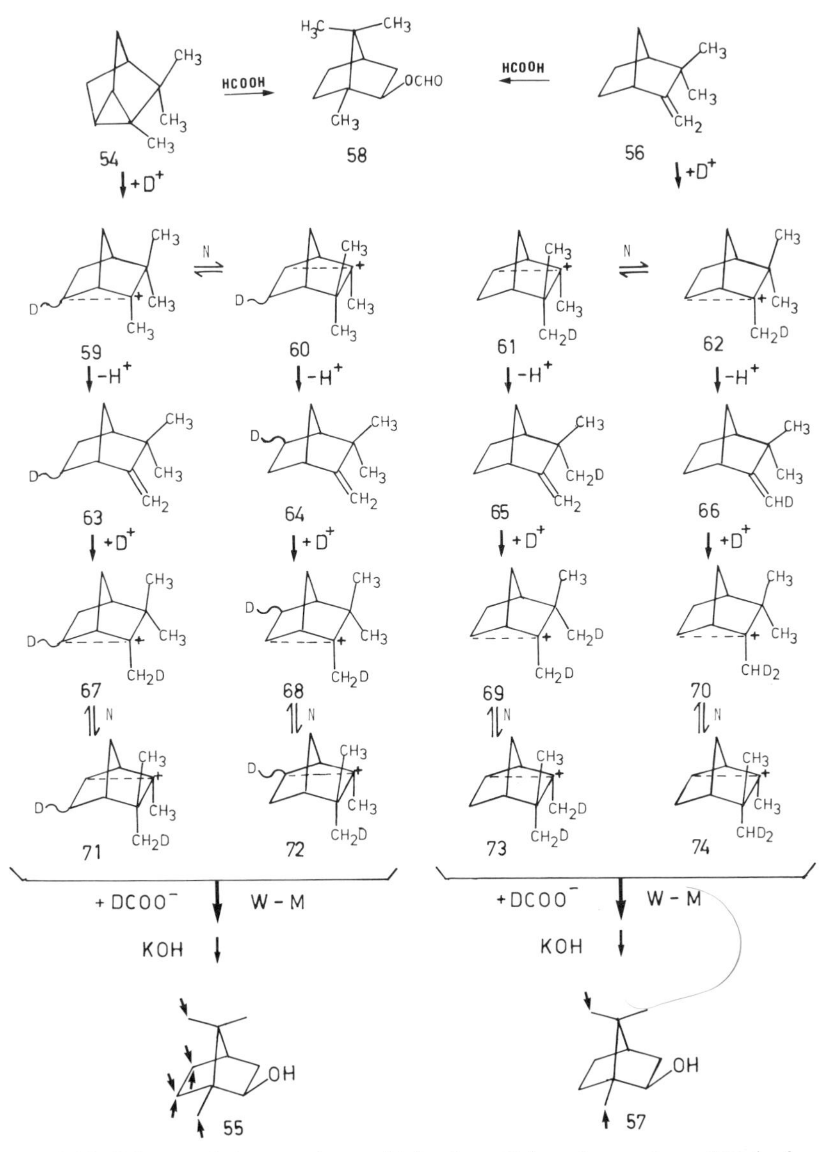

Figure 4-18. Scheme of the reactions of tricyclene (**54**) and camphene (**56**) in formic acid (HCOOH) and in dideuteroformic acid (DCOOD). The arrows on the formulas of saponification products **55** and **57** indicate the positions where deuterium has been incorporated as a result of the main reaction scheme. (W-M = Wagner Meerwein rearrangement.)

The starting reaction of tricyclene (**54**) is deuteronation to the corner of the cyclopropane ring (C-6), analogous to the reaction of nortricyclene (**44**, Figure 4-15) and 1-methylnortricyclene (**49**, Figure 4-16). In contrast, a deuteron adds to camphene (**56**) to the double bond carbon (C-10). The camphene cations **59** and **62** initially formed are in rapid equilibrium exchange with ions **60** and **61**, respectively. This exchange involves Nametkin rearrangement (N), where the methyl group at the exo-3 position very easily moves to the exo-2 position. This *exo*-methyl 3–2 migration has been shown to be the dominant phenomenon in the acid-catalyzed racemization of camphene (**56**) by studies where, eg, ^{14}C labeling and ^{13}C NMR were used.[104] Another phenomenon that takes place by exo attack is Wagner-Meerwein rearrangement of the tertiary ion to the secondary exo product (isobornyl formate **58** or its deuterated analogs). Before this product-forming Wagner-Meerwein reaction occurs, the tertiary ions **59**–**62** may give away protons to form deuterated camphenes **63**–**66**, which then may react with a new molecule of dideuteroformic acid. This is evident from the fact that methyl groups 9 and 10 in **55** also are deuterated. Series of equilibrating ions (**59**–**62** and **67**–**74**, etc) serve as precursors of the differently deuterated esters that are formed from them by Wagner-Meerwein rearrangement with attack of the formate anion from the exo direction.

Concluding Remarks

Use of lanthanide shift reagents greatly expands the capacity of ^{1}H NMR to determine the stereostructures, conformations, and steric positions of the label atoms in organic molecules that contain heteroatoms with coordinating electron pairs. Combined use of this methodology with ^{13}C and high-field NMR and NMR of other nuclei with or without LSR is recommended for complex cases.

References

1. Reuben, J.; Fiat, T. *J. Chem. Phys.*, **1969**, *51*, 4909.
2. Golding, R. M.; Halton, M. P. *Australian J. Chem.*, **1972**, *25*, 2577.
3. Bleaney, B. *J. Magn. Reson.*, **1972**, *8*, 91.
4. Golding, R. M.; Pyykkö, P. *Mol. Phys.* **1973**, *26*, 1389.
5. McConnell, H. M.; Robertson, R. E. *J. Chem. Phys.*, **1958**, *29*, 1361.
6. Farid, S.; Ateya, A.; Maggio, M. *Chem. Commun.* **1971**, 1285.
7. Häkli, H. "Structure Analysis and Molecular Dynamics of Cyclic Compounds by

Shift Reagent NMR"; Research Report No. 7, Department of Chemistry, University of Jyväskylä, Finland 1979; Ph.D. dissertation.
8. Reuben, J. "Paramagnetic Lanthanide Shift Reagents in NMR Spectroscopy: Principles, Methodology and Applications"; Pergamon Press: Oxford, 1973.
9. Paasivirta, J. *Suomen Kemistilehti*, **1971**, *B44*, 131.
10. Demarco, P. V.; Elzey, T. K.; Lewis, R. B.; Wenkert, E. *J. Am. Chem. Soc.*, **1970**, *92*, 5734, 5737.
11. Armitage, I. M.; Hall, L. D.; Marshall, A. G.; Werbelow, L. G. In "Nuclear Magnetic Resonance Shift Reagents"; Sievers, R. E., Ed.; Academic Press: New York, 1973; p. 313.
12. Shapiro, B. L.; Johnston, M. D., Jr. *J. Am. Chem. Soc.*, **1972**, *94*, 8185.
13. Shapiro, B. L.; Johnston, M. D., Jr.; Towns, R. L. R.; Godwin, A. D.; Pearce, H. L.; Proulx, T. W.; Shapiro, M. J. In "Nuclear Magnetic Resonance Shift Reagents"; Sievers, R. E., Ed., Academic Press: New York, 1973; p. 227.
14. Talvitie, A. "Structure Determination of Some Sesquiterpenoids by Shift Reagent NMR"; Research Report No. 6, Department of Chemistry, University of Jyväskylä, 1979; Ph.D. dissertation.
15. Paasivirta, J. *Suomen Kemistilehti*, **1973**, *B46*, 159, 162.
16. Paasivirta, J.; Häkli, H.; Forsén, K. *Finn. Chem. Lett.*, **1974**, 165.
17. Paasivirta, J.; Häkli, H.; Widen, K-G, *Org. Magn. Reson.*, **1974**, *6*, 380.
18. Johnston, M. D., Jr.; Shapiro, B. L.; Shapiro, M. J.; Proulx, T. W.; Godwin, A. D.; Pearce, H. L. *J. Am. Chem. Soc.*, **1975**, *97*, 542.
19. Ajisaka, K.; Kainosho, M. *J. Am. Chem. Soc.*, **1975**, *97*, 330.
20. Gansow, O. A.; Loeffer, P. A.; Davis, R. E.; Lenkinski, R. E.; Willcott, M. R. *J. Am. Chem. Soc.* **1976**, *98*, 4250.
21. Brown, H. C. "The Nonclassical Ion Problem," with comments by P. von R. Schleyer; Plenum Press: New York, 1977.
22. Toivonen, N. J. In "6th Nordic Chemistry Congress"; Chemical Society of Sweden, Lund, 1947; Abstract p. 276.
23. Toivonen, N. J.; Hirsjärvi, P.; Melaja, A.; Kainulainen, A.; Halonen, A.; Pulkkinen, E. *Acta Chem. Scand.*, **1949**, *3*, 991.
24. Hirsjärvi, P. *Ann. Acad. Sci. Fenn.*, *AII*, *Chem.*, **1952**, *45*.
25. Hirsjärvi, P. *Suomen Kemistilehti*, **1956**, *B29*, 138.
26. Pulkkinen, E. *Ann. Acad. Sci. Fenn.*, *AII*, *Chem.*, **1956**, *74*.
27. Musher, J. I. *Mol. Phys.* **1963**, *6*, 93.
28. Äyräs, P.; Paasivirta, J. *Suomen Kemistilehti*, **1969**, *B42*, 61.
29. Abraham, R. J.; Bernstein, H. J. *Can. J. Chem.*, **1961**, *39*, 216.
30. Paasivirta, J.; Mälkönen, P. J. *Suomen Kemistilehti*, **1971**, *B44*, 230.
31. Toivonen, N. J.; Helske, M.; Mälkönen, P. J.; Siltanen, E.; Tuovinen, M.; Ojala, K. Suomen Kemistilehti, **1954**, *A27*, 348.
32. Toivonen, N. J.; Siltanen, E.; Ojala, K. *Ann. Acad. Sci. Fenn.*, *AII Chem.*, **1955**, *64*.
33. Toivonen, N. J. "XIV Internationaler Kongress für Reine und Angewandte Chemie"; Referatendband; Zürich, 1955; p. 45.
34. Toivonen, N. J. In "9th Nordic Chemistry Congress"; Chemical Society of Denmark; Plenumsforedrag: Aarhus, 1956; p. 117.
35. Mälkönen, P. J. *Suomen Kemistilehti*, **1969**, *B42*, 230.
36. Mälkönen, P. J. *Suomen Kemistilehti*, **1971**, *B44*, 12.
37. Paasivirta, J.; Mälkönen, P. J. *Suomen Kemistilehti*, **1971**, *B44*, 283.
38. Mälkönen, P. J. *Suomen Kemistilehti*, **1962**, *B35*, 250.
39. Mälkönen, P. J. *Ann. Acad. Sci. Fenn.*, *AII*, *Chem.*, **1964**, *128*.
40. (a) Gerber, N. N. *J. Antibiot. Tokyo*, **1969**, *22*, 508; (b) *J. Chem. Ecol.*, **1977**, *3*, 475.
41. Medsker, L. L., Jenkins, D.; Thomas, J. F., Koch, C. *Environ. Sci. Technol.*, **1969**, *3*, 476.
42. Paasivirta, J. *Suomen Kemistilehti*, **1971**, *B44*, 135.

43. Paasivirta, J. *Ann. Acad. Sci. Fenn., AII, Chem.*, **1962**, *116*.
44. Paasivirta, J. *Suomen Kemistilehti*, **1969**, *B42*, 37.
45. Tai, J. C.; Allinger, N. L. *J. Am. Chem. Soc.*, **1966**, *88*, 2179.
46. Corey, E. J.; Sneen, R. A. *J. Am. Chem. Soc.*, **1955**, *77*, 2505.
47. Stilbs, P. *Chem. Scripta*, **1975**, *7*, 59.
48. von Rudloff, E. *Can. J. Chem.*, **1963**, *41*, 1737.
49. von Schantz, M.; Forsén K. *Farm. Aikak.*, **1971**, *80*, 122.
50. Matsuo, A.; Uchio, Y.; Nakayama, M.; Matsubara, Y.; Hayashi, S. *Tetrahedron Lett.*, **1974**, 4219.
51. Chiang, J. F., Willcox, C. F., Jr.; Bauer, S. H. *J. Am. Chem. Soc.*, **1968**, *90*, 3149.
52. Laatikainen, R. "Computerized NMR: Developments and Applications of Analysis and Interpretation Methods of High-Resolution Spectra"; Ph.D. dissertation, Department of Chemistry, University, of Jyväskylä, 1979.
53. Whittaker, E. T.; Robinson, G. "The Calculus of Observations"; Blackie and Son: London, 1929; p. 209.
54. Paasivirta, J.; Häkli, H.; Widen, K-G. Unpublished results.
55. Willcott, M. R.; Davis, R. E.; Holder, R. W. *J. Org. Chem.*, **1975**, *40*, 1952.
56. Abraham, R. J.; Coppell, S. M.; Ramage, R. *Org. Magn, Reson.*, **1974**, *6*, 658.
57. Reuben, J. *J. Magn. Reson.*, **1973**, *11*, 103.
58. Bell, H. M. *Org. Magn. Reson.*, **1975**, *7*, 240.
59. Mariano, P. S.; McElroy, R. *Tetrahedron Lett.* **1972**, 5305.
60. Williamson, K. L.; Clutter, D. R.; Emch, R.; Alexander, M.; Burroughs, A. E.; Chua, C.; Bogel, M. E. *J. Am. Chem. Soc.*, **1974**, *96*, 1471.
61. Lewis, R. B., Wenkert, E. In "Nuclear Magnetic Resonance Shift Reagents"; Sievers, R. E., Ed.; Academic Press: New York, 1973; p. 99.
62. Mohyla, I.; Ksandr, Z.; Hájek, M.; Vodicka, L. *Coll. Czech. Chem. Commun.*, **1974**, *39*, 2935.
63. Duddeck, H.; Dietrich, W. *Tetrahedron Lett.*, **1975**, 2925.
64. Groves, J. T.; Van Der Puy, M. *Tetrahedron Lett.*, **1975**, 1949.
65. Green, M. M.; Cook, R. J. *J. Am. Chem. Soc.*, **1969**, *91*, 2129.
66. Borg-Karlsson, A-K.; Norin, T.; Talvitie, A. *Tetrahedron*, **1981**, *37*, 425.
67. Talvitie, A.; Borg-Karlsson, A-K. *Finn. Chem. Lett.*, **1979**, 93.
68. Talvitie, A.; Norin, T.; Strömberg, S.; Weber, M. *Finn. Chem. Lett.*, **1979**, 149.
69. Borg-Karlsson, A-K.; Norin, T.; Wijekoon, W. M. D.; Talvitie, A. *Finn. Chem. Lett.*, **1979**, 151.
70. Talvitie, A.; Paasivirta, J.; Widen, K-G, *Finn. Chem. Lett.*, **1977**, 197.
71. Lenkinski, R. E.; Reuben, J. *J. Am. Chem. Soc.*, **1976**, *98*, 4065.
72. Finocciaro, P.; Recca, A.; Meravigana, P.; Montaudo, G. *Tetrahedron*, **1974**, *30*, 4159.
73. Korvola, J.; Häkli, H.; Paasivirta, J. *Finn. Chem. Lett.*, **1974**, 16.
74. Winstein, S.; Trifan, D. S. *J. Am. Chem. Soc.*, **1949**, *71*, 2953.
75. Olah, G. A.; Commeyras, A.; Lui, C. Y. *J. Am. Chem. Soc.*, **1968**, *90*, 3882.
76. Olah, G. A.; White, A. M. *J. Am. Chem. Soc.*, **1969**, *91*, 3956.
77. Olah, G. A.; White, A. M.; DeMember, J. M.; Commeyras, A.; Lui, C. Y. *J. Am. Chem. Soc.*, **1970**, *92*, 4627.
78. Brown, H. C. "The Transition State", Chem. Soc. Spec. Publ. No 16; London 1962; 140, 174.
79. Brown, H. C. *Chem. Eng. News*, **1967**, 87.
80. Brown, H. C.; Kawakami, J. H.; Liu, K-T. *J. Am. Chem. Soc.*, **1970**, *92*, 5536.
81. Paasivirta, J. *Acta Chem. Scand.*, **1973**, *27*, 374.
82. Pulkkinen, E. *Suomen Kemistilehti*, **1954**, *A27*, 26.
83. Kleinfelter, D. C.; von R. Schleyer, P. *Org. Synth.*, **1963**, *42*, 79.
84. Paasivirta, J. *Acta Chem. Scand.*, **1968**, *22*, 2200.

85. Paasivirta, J.; Knihtilä, H.; Mäkelä, R.; Arkko, A.; Suontamo, R.; Surma-Aho, K. Poster presentation, conference on the Chemistry of Carbocations; Bangor, N. Wales, Sept. 7–11, 1981, and unpublished results.
86. Brown, H. C.; Liu, K-T. *J. Am. Chem. Soc.*, **1967**, *89*, 3900.
87. Brown, H. C.; Liu, K-T. *J. Am. Chem. Soc.*, **1975**, *97*, 600.
88. Brown, H. C.; Kawakami, J. H. *J. Am. Chem. Soc.*, **1975**, *87*, 5521.
89. Liu, K-T. *Tetrahedron Lett.*, **1977**, 1207.
90. Werstiuk, N. H.; Kadai, T. *Chem. Commun.*, **1971**, 1349.
91. Brown, J. M.; McIvor, M. C. *Chem. Commun.*, **1969**, 238.
92. Nickon, A.; Hammons, J. H. *J. Am. Chem. Soc.*, **1964**, *86*, 3322.
93. Liu, K-T. *Tetrahedron Lett.* **1978**, 1129.
94. Nordlander, J. E.; Owuor, P. O.; Haky, J. E. *J. Am. Chem. Soc.*, **1979**, *101*, 1288.
95. Grob, C. A. *Angew. Chem.*, **1982**, *94*, 87.
96. Fischer, W.; Grob, C. A.; Hanreich, R.; von Sprecher, G.; Waldner, A. *Helv. Chim. Acta*, **1981**, *64*, 2298.
97. Grob, C. A.; Günther, B.; Hanreich, A. *Helv. Chim. Acta*, **1981**, *64*, 2312.
98. Paasivirta, J. *Justus Liebigs Ann. Chem.*, **1965**, *686*, 1.
99. Olah, G. A.; DeMember, J. R.; Lui, C. Y.; White, A. M. J. *J. Am. Chem. Soc.*, **1969**, *91*, 3958.
100. Grob, C. A.; Waldner, A. *Tetrahedron Lett.*, **1980**, *21*, 4429, 4433.
101. Paasivirta, J.; Kuorikoski, S-L.; Laasonen, T. *Finn. Chem. Lett.*, **1978**, 42.
102. Meérwein, H.; van Emster, K. *Chem. Ber.*, **1922**, *55*, 2500.
103. Nevell, T. P.; deSalas, E.; Wilson, C. L. *J. Chem. Soc.*, **1939**, 1188.
104. David, C. W.; Everling, B. W.; Kilian, R. J.; Stothers, J. B.; Vaughan, W. R. *J. Am. Chem. Soc.*, **1973**, *95*, 1265.

Appendix 1

```
138*LANTA(1).MAIN
     1      C
     2      C
     3      C           LANTA IS A PROGRAM THAT OPTIMIZES THE POSITION OF THE LN ATOM
     4      C           IN A MOLECULE COMPLEX FOR DIPOLAR MODEL
     5      C           ONE CAN OPTIMIZE THE PLACE OF LN ATOM (VARIABLES 1-3) AND
     6      C           PLACE OF THE HETEROATOM (VARIABLES 4-6) AND ONE CAN DEFINE
     7      C           THE DISTANCE LN-HETEROATOM AND THE CORRESPONDING ANGLES
     8      C           TO BE CONSTANT DURING THE ITERATION
     9      C
    10      C           PROGRAMMING IS BASED ON THE ITERATION OF NEWTON AND RAPHSON:
    11      C           TO ASSURE THE CONVERGENCE THE MAXIMAL STEP IN A CHANGE OF A
    12      C           COORDINATE IS SPECIFIED TO BE 0.2 UNITS OR LESS
    13      C
    14      C           THE PROGRAM IS WRITTEN BY REINO LAATIKAINEN / JYVÄSKYLÄ UNIVERSITY
    15      C           THIS VERSION IS FIT TO UNIVAC 1100
    16      C
    17      C
    18      C
    19            DIMENSION MITA(20),IFX(20),PRO(20),ERR(3,10),CONSTR(5)
    20            COMMON /COORDI/Z(10),C(3),H(20,3),EUH2(20),CS(20)
    21           +/OBSED/B(20),B2(20),B3(20),CORR(10),DERI(10),OBS(20)
    22           +/MATRIX/DH(20,10),X(10,10),DEV(20),Y(10)
    23            WRITE(6,111)
    24      111   FORMAT(///" PROGRAM LANTA FOR ANALYSIS OF LN SHIFTS IN"
    25           +          /" NMR CHEMICAL SHIFT"//)
    26            NOX=6
    27            YY=0.
    28      C     INPUT VALUES ARE READ IN
    29            READ(27,2)ITE,Z(1),Z(2),Z(3)
    30            READ(27,2)JJ,Z(4),Z(5),Z(6)
    31            HE1=Z(4)
    32            HE2=Z(5)
    33            HE3=Z(6)
    34            DELMAX=-9999.9
    35            READ(27,2)JJ,C(1),C(2),C(3)
    36            IF(OBS(L).GT.DELMAX)DELMAX=OBS(L)
    37            DO 10 L=1,20
    38            READ(27,2,END=299)JJ,H(L,1),H(L,2),H(L,3),OBS(L)
    39         10 YY=YY+OBS(L)
    40        299 NH=L-1
    41         16 RMS=99999.9
    42            CONSTA=1.0
    43            LUKU=0
    44            ITEX=0
    45            NCON=0
    46            IWRIT=0
    47            WRITE(6,19)
    48            READ(5,20)CONSTR(1)
    49            IF(CONSTR(1).LT.0.0001)GO TO 30
    50            READ(5,20)HELN
    51            NCON=NCON+1
    52         30 WRITE(6,18)
    53            READ(5,20)CONSTR(2)
    54            IF(CONSTR(2).LT.0.0001)GO TO 31
    55            READ(5,20)HESU
    56            NCON=NCON+1
    57         31 WRITE(6,33)
    58            READ(5,20)CONSTR(3)
    59            IF(CONSTR(3).LT.0.0001)GO TO 32
    60            READ(5,20)FII
    61            NCON=NCON+1
    62         20 FORMAT(F10.5)
    63         32 WRITE(6,7)
```

```
 64            READ(5,112)(IFX(K),K = 1,10)
 65       112  FORMAT(10I1)
 66            NDF = NH + NCON - 2
 67            DO 349 L = 1,10
 68       349  MITA(L) = 0
 69            IFNOX = 0
 70            DO 350 L = 1,10
 71            J = IFX(L)
 72            IF(J.EQ.O)GO TO 350
 73            MITA(J) = 1
 74            NDF = NDF - 1
 75            IF(J.GT.3)IFNOX = 1
 76       350  CONTINUE
 77            IF(NDF.LT.1)NDF = 1.0
 78            SUBROUTINE COMPU FORMS THE JACOBIAN
 79        11  CALL COMPU(NH,IFNOX,CHE2,EUC2,EUHE2)
 80            CHE = SQRT(CHE2)
 81            EUHE = SQRT(EUHE2)
 82            IF(YY.GT.0.001)GO TO 23
 83            IF OBSERVED SHIFTS ARE NOT GIVEN, PROGRAM ACTS AS
 84            SIMULATION PROGRAM
 85            DELMAX = -9999.9
 86            DO 24 L = 1,NH
 87            IF(B(L).GT.DELMAX)DELMAX = B(L)
 88        24  CONTINUE
 89            DO 25 L = 1,NH
 90            OBS(L) = 100.0*B(L)/DELMAX
 91        25  YY = YY + OBS(L)
 92            IWRIT = 1
 93    C       LEAST SQUARE FITTING IS CARRIED
 94        23  CALL SQUARE(NH,B,B2,OBS,YY,TAN,BB)
 95            SQ = 0.0
 96            DO 60 L = 1,NH
 97            DEV(L) = B2(L) - OBS(L)
 98        60  SQ = SQ + DEV(L)*DEV(L)
 99            RMS2 = RMS
100            RMS = SQRT(SQ/NDF)
101            DO 61 L = 1,6
102            IF(MITAC(L).EQ.O)GO TO 61
103            DO 62 K = 1,NH
104        62  B3(K) = B(K) + DH(K,L)*0.01
105            CALL SQUARE(NH,B3,B3,OBS,YY,TN,BBB)
106            DO 63 K = 1,NH
107        63  DH(K,L) = (B3(K) - B2(K))/0.01
108        61  CONTINUE
109    C       CONSTRAINS ARE APPLIED
110            SQCON = 0.0
111            NOBS = NH
112            NOBS = NOBS + 1
113            DO 50 L = 1,3
114            L2 = L + 3
115            DH(NOBS,L) = (Z(L) - Z(L2))/EUHE
116        50  DH(NOBS,L2) = -DH(NOBS,L)
117            IF(CONSTR(1).GT.0.)SQCON = SQCON + (EUHE - HELN)**2
118            DEV(NOBS) = (EUHE - HELN)*CONSTR(1)
119            NOBS = NOBS + 1
120            DO 550 L = 1,3
121            LS = L + 3
122            DH(NOBS,L) = 0.0
123       550  DH(NOBS,L2) = (Z(L2) - C(L))/CHE
124            IF(CONSTR(2).GT.O.)SQCON = SQCON + (HESU - CHE)**2
125            DEV(NOBS) = (CHE - HESU)*CONSTR(2)
126            NOBS = NOBS + 1
127            CALL DANGLE(NOBS,EUHE2,CHE2,EUC2,FII2,1)
128            IF(CONSTR(3).GT.O.)SQCON = SQCON + (FII2 - FII)**2
129            DEV(NOBS) = (FII2 - FII)*CONSTR(3)
130    C       RRMS - VALUES ARE COMPUTED AND CONVERGENCE TESTED
131       253  RMSCON = SQRT(SQCON/NCON)
132            WRITE(6,3)RMS,RMSCON,CONSTA
133            TESTI = ABS(RMS2 - RMS)/(RMS + 0.1)
134            LUKU = LUKU + 1
135            IF(TESTI.GT.0.001.AND.LUKU.LT.ITEX)GO TO 1111
136            WRITE(6,4)
137            READ(5,1)ITEX
```

```
138              IF(ITEX.GT.20)ITEX=20
139              LUKU=0
140     C        NORMA CREATES THE NORMAL EQUATIONS
141        1111  DO 70 L=1,3
142              L1=NH+L
143              DO 71 K=1,NOX
144              ERR(L,K)=DH(L1,K)
145          71  DH(L1,K)=CONSTR(L)*DH(L1,K)
146          70  CONTINUE
147              CALL NORMA(NOBS,NOX)
148              DO 123 L=1,NOX
149              IF(MITA(L).NE.O)GO TO 123
150              DO 1234 K=1,NOX
151              X(L,K)=0.0
152        1234  X(K,L)=0.0
153              X(L,L)=1000.0
154              Y(L)=0.0
155         123  CONTINUE
156     C        INVERT COMPUTES INVERSE OF X-MATRIX
157     C        CALL MATOUT(X,NOX)
158              CALL INVERT(X,NOX)
159     C        CALL MATOUT(X,NOX)
160     C        CORRECTIONS TO VARIABLES ARE COMPUTED, DAMPING IS
161     C        APPLIED IF STEP>0.2 UNITS
162              IF(ITEX.EQ.O)GO TO 154
163              DO 51 L=1,NOX
164              CORR(L)=0.00
165              DO 51 K=1,NOX
166          51  CORR(L)=CORR(L)+X(L,K)*Y(K)
167              DO 53 L=1,NOX
168          53  PRO(L)=ABS(CORR(L)/0.2)
169              CONSTA=1.0
170              DO 54 L=1,NOX
171          54  IF(PRO(L).GT.CONSTA)CONSTA=PRO(L)
172              DO 55   L=1,NOX
173              CORR(L)=CORR(L)/CONSTA
174              IF(MITA(L).EQ.O)CORR(L)=0.0
175          55  CONTINUE
176              DO 52 L=1,NOX
177          52  Z(L)=Z(L)-CORR(L)
178              GO TO 11
179     C        OUTPUT OF ITERATED VALUES OF VARIABLES
180         154  WRITE(6,5)
181              DF=NDF*1.0
182     C        MDSTI IS IMSL-SUBROUTINE: COMPUTES STUDENT-T VALUE
183     C                  ****
184     C        FOR RISK 0.1 (OR ANY)
185              CALL MDSTI(0.1,DF,STUDNT,IER)
186              DO 170 L=1,NOX
187              DERI(L)=STUDNT*RMS*SQRT(ABS(X(L,L)))
188              IF(MITA(L).EQ.O)DERI(L)=0.0
189         170  CONTINUE
190              WRITE(6,6)(Z(L),DERI(L),L=1,3)
191              WRITE(6,8)
192              WRITE(6,6)(Z(L),DERI(L),L=4,6)
193              WRITE(6,9)(J,J=1,NOX)
194              DO 40 L=1,NOX
195              DO 41 K=1,NOX
196          41  CORR(K)=X(L,K)/SQRT(ABS(X(L,L)*X(K,K)))
197          40  WRITE(6,2)L,(CORR(J),J=1,NOX)
198     C        OUTPUT OF SOME STRUCTURAL PARAMETERS, ETC.
199              EUHE2=(Z(4)-Z(1))**2+(Z(5)-Z(2))**2+(Z(6)-Z(3))**2
200              EUHX=(Z(1)-HE1)**2+(Z(2)-HE2)**2+(Z(3)-HE3)**2
201              HEHE2=(Z(4)-HE1)**2+(Z(5)-HE2)**2+(Z(6)-HE3)**2
202              COME=(EUHX+EUHE2-HEHE2)/(2*SQRT(EUHX*EUHE2))
203              OMEG=180*ATAN(SQRT(1-COME**2)/COME)/3.14159
204              CALL DANGLE(NOBS,EUHE2,CHE2,EUC2,FII2,0)
205              CORR(4)=0.0
206              DO 80 L=1,NOX
207              Z(L)=Z(L)+DERI(L)
208              EUHE22(Z(4)-Z(1))**2+(Z(5)-Z(2))**2+(Z(6)-Z(3))**2
209              EUHX2=(Z(1)-HE1)**2+(Z(2)-HE2)**2+(Z(3)-HE3)**2
210              HEHE22=(Z(4)-HE1)**2+(Z(5)-HE2)**2+(Z(6)-HE3)**2
211              COME2=(EUHX2+EUHE22-HEHE22)/(2*SQRT(EUHX2*EUHE22))
```

```
212           OMEG2=180*ATAN(SQRT(1-COME2**2)/COME2)3.14159
213           CORR(4)=CORR(4)+(OMEG-OMEG2)**2
214           Z(L)=Z(L)-DERI(L)
215        80 CONTINUE
216           CORR(4)=SQRT(CORR(4))
217           DO 75 L=1,3
218           CORR(L)=0.0
219           DO 76 K=1,NOX
220        76 CORR(L)=CORR(L)+(DERI(K)*ERR(L,K))**2
221        75 CORR(L)=SQRT(CORR(L))
222           WRITE(6,605)EUHE,CORR(1),CHE,CORR(2),FII2,CORR(3),OMEG
223          1,CORR(4)
224           WRITE(6,606)TAN,BB
225           AVERG=YY/NH
226           SQTOT=0.0
227           DO 650 L=1,NH
228       650 SQTOT=SQTOT+(OBS(L)-AVERG)**2
229           R2=1.0-SQ/SQTOT
230           R=SQRT(R2)
231           WRITE(6,656)R,R2
232           WRITE(6,17)
233           READ(5,1)MITTEE
234           IF(MITTEE)16,15,16
235        15 WRITE(6,601)
236           DO 600 L=1,NH
237           EH=SQRT(EUH2(L))
238           B(L)=TAN*B(L)+BB
239           ANGLE=180*ATAN(SQRT(1-CS(L))/SQRT(CS(L)))/3.14159
240       600 WRITE(6,2)L,B(L),OBS(L),DEV(L),EH,ANGLE
241           WRITE(6,700)
242           DO 173 L=1,NH
243       173 WRITE(6,2)L,(DH(L,K),K=1,NOX)
244           DO 175 L=1,3
245       175 WRITE(6,2)L,(ERR(L,K),K=1,NOX)
246           IF(IWRIT.EQ.O)GO TO 999
247   C       OUTPUT OF SIMULATION RESULT (IF OBSERVED SHIFTS ARE NOT
248   C       GIVEN IN INPUT): REPLACES THE INPUT FILE
249           REWIND 27
250           J=111
251           WRITE(27,2)J,Z(1),Z(2),Z(3)
252           WRITE(27,2)J,Z(4),Z(5),Z(6)
253           WRITE(27,2)J,C(1),C(2),C(3)
254           DO 998 L=1,NH
255       998 WRITE(27,2)L,(H(L,K),K=1,3),OBS(L)
256       605 FORMAT(/' LANTHANIDE-OXYGEN',2F10.4/' OXYGEN-CARBON',F14.4,F10.4
257          +' FII-ANGLE',8X,2F10.4/' OMEGA-ANGLE',6X,2F10.4/)
258       606 FORMAT(' REGRESSION EQUATION:'/' D(H)=',F8.1,'*(1-3COS2 X)/R3'
259          1           ' +',F8.1/)
260       656 FORMAT(' CORRELATION COEFFICIENT R =',F10.6,' (R2 =',F10.6,')'/)
261       601 FORMAT(//'     H          CALC          OBS          DIFF.     EU-H  '
262          +' H-LN-X'/)
263       700 FORMAT(/' D(SHIFT)/D(VARIABLE)'/)
264        19 FORMAT(' IF YOU WANT CONSTRAIN LN-HET.ATOM DISTANCE, GIVE (F8.3)
265          +' NON-ZERO VALUE FOR'/' F-CONSTANT AND DISTANCE IN ANGSTROMS:')
266        18 FORMAT(' SAME FOR HET.ATOM-SUBSTRATE DISTANCE....')
267        33 FORMAT(' SAME FOR FII-ANGLE,...DEFINE ANGLE IN DEGREES')
268        17 FORMAT(' WRITE 1 IF YOU WANT GO ON WITH CALCULATION!')
269         4 FORMAT(/' GIVE (12) NO. ITERATIONS:')
270         5 FORMAT(//' LN-ATOM; COORDINATES AND 90% CONFIDENCE LIMITS')
271         7 FORMAT(' DEFINE (1011) VARIABLES TO BE ITERATED: LN-ATOM'
272          1', VARIABLES 1-3'/' HE-ATOM, VARIABLES 4-6')
273         8 FORMAT(' HETERO-ATOM:')
274         9 FORMAT(' CORRELATIONS OF VARIABLES:'/1X,10I10)
275         6 FORMAT(' X',2F10.4/' Y',2F10.4/' Z',2F10.4/)
276         1 FORMAT(I2)
277         2 FORMAT(I3,8F10.3)
278         3 FORMAT(' RRMS VALUES:',2F8.3,' (DAMPING FACTOR' ,F8.2,')')
279       999 WRITE(6,900)
280       900 FORMAT(///'     F I N     !!!  '///)
281           STOP
282           END
283           SUBROUTINE COMPU(NOH,IFNOX,CHE2,EUC2,EUHE2)
284   C       (1-3COS2 X)/R3-VALUES AND DIFFERENTIALS
285   C       D((1-3COS...)/D(VARIABLE ARE COMPUTED
```

```
286          COMMON /COORDI/Z(10),C(3),H(20,3),EUH2(20),CS(20)
287         +/OBSED/B(20),B2(20),CORR(10),DERI(10),OBS(20)
288         +/MATRIX/DH(20,10),X(10,10),DEV(20),Y(10)
289          EUHE2=(Z(4)-Z(1))**2+(Z(5)-Z(2))**2+(Z(6)-Z(3))**2
290          CHE 2=(C(1)-Z(4))**2+(C(2)-Z(5))**2+(C(3)-Z(6))**2
291          EUC2=(C(1)-Z(1))**2+(C(2)-Z(2))**2+(C(3)-Z(3))**2
292          DO 1 L=1,NOH
293          EUH2(L)=(H(L,1)-Z(1))**2+(H(L,2)-Z(2))**2+(H(L,3)-Z(3))**2
294          HHE2=(H(L,1)-Z(4))**2+(H(L,2)-Z(5))**2+(H(L,3)-Z(6))**2
295          G=3*(EUH2(L)+EUHE2-HHE2)**2
296          U=4*(EUH2(L)*EUHE2)
297          V=EUH2(L)**1.50
298          CS(L)=G/(3*U)
299          B(L)=(G/U-1)/V
300          DO 1 N=1,3
301          M=N+3
302          GD=12.0*(EUH2(L)+EUHE2-HHE2)*(2*Z(N)-H(L,N)-Z(M))
303          VD=3.0*SQRT(EUH2(L))*(Z(N)-H(L,N))
304          UD=8*(EUH2(L)*(Z(N)-Z(M)))+8*(EUHE2*(Z(N)-H(L,N)))
305          DH(L,N)=(GD-UD)/(V*U)-(V*UD+VD*U)*(G-U)/(U*V)**2
306          IF(IFNOX.EQ.O)GO TO 1
307          GDO=12*(EUH2(L)+EUHE2-HHE2)*(H(L,N)-Z(N))
308          UDO=8*EUH2(L)*(Z(M)-Z(N))
309          DH(L,M)=(1/V)*(GDO*U-UDO*G)/(U**2)
310        1 CONTINUE
311          RETURN
312          END
313          SUBROUTINE SQUARE(NOBSD,B,B2,OBS,YY,TAN,BB)
314   C      LEAST SQUARE FITTING
315          DIMENSION B(20),B2(20),OBS(20)
316          XX=0.00
317          XY=0.00
318          X2=0.00
319          DO 1 L=1,NOBSD
320          XX=XX+B(L)
321          X2=X2+B(L)**2
322        1 XY=XY+OBS(L)*B(L)
323          TAN=(NOBSD*XY-XX*YY)/(NOBSD*X2-XX*XX)
324          BB=(YY*X2-XY*XX)/(NOBSD*X2-XX*XX)
325          DO 2 L=1,NOBSD
326        2 B2(L)=TAN*B(L)+BB
327          RETURN
328          END
329          SUBROUTINE DANGLE(NOBS,R,S,T,FII2,IFDIF)
330   C     DIFFERENTIALS D(FII)/D(VARIABLE) ARE COMPUTED, FII-
331   C     ANGLES AND COMPUTED
332          COMMON /COORDI/Z(10),C(3),H(20,3),EUH2(20),CS(20)
333         +/MATRIX/DH(20,10),X(10,10),DEV(20),Y(10)
334          RS=S*R
335          RST=S+R-T
336          G=RST/(2*SQRT(RS))
337          PIICON=180/3.14159
338          FII2=ARCOS(G)
339          FII2=PIICON*FII2
340          IF(IFDIF.EQ.OO.O)RETURN
341          DO 1 L=1,3
342          L2=L+3
343          DRST=C(L)-Z(L2)
344          DRS=-S*(Z(L)-Z(L2))
345          DG=DRS*G/RS+DRST/SQRT(RS)
346          DH(NOBS,L)=-PIICON*DG/SQRT(1-G**2)
347          DRS=-(S*(Z(L2)-Z(L))+R*(Z(L2)-C(L)))
348          DRST=2*Z(L2)-Z(L)-C(L)
349          DG=DRS*G/RS + DRST/SQRT(RS)
350          DH(NOBS,L2)=-PIICON*DG/SQRT(1-G**2)
351        1 CONTINUE
352          RETURN
353          END
354          SUBROUTINE NORMA(NOBSD,NOX)
355   C      NORMAL EQUATIONS ARE FORMED FROM D-MATRIX
356          COMMON /MATRIX/D(20,10),X(10,10),DEV(20),Y(10)
357          DO 210 NS1=1,NOX
358          DO 206 NS2=NS1,NOX
359          X(NS1,NS2)=0.00
```

```
360            DO 205 LEQ = 1,NOBSD
361       205  X(NS1,NS2) = X(NS1,NS2) + D(LEQ,NS1)*D(LEQ,NS2)
362       206  X(NS2,NS1) = X(NS1,NS2)
363            Y(NS1) = 0.0
364            DO 210 LEQ = 1,NOBSD
365       210  Y(NS1) = Y(NS1) + D(LEQ,NS1)*DEV(LEQ)
366            RETURN
367            END
368            SUBROUTINE INVERT(A,N)
369            DIMENSION A(10,10),B(10),C(10),LZ(10)
370            DO 10 J = 1,N
371        10  LZ(J) = J
372            DO 20 I = 1,N
373            K = 1
374            Y = A(I,I)
375            L = I - 1
376            LP = I + 1
377            IF(N - LP)14,11,11
378        11  DO 13 J = LP,N
379            W = A(I,J)
380            IF(ABS(W) - ABS(Y))13,13,12
381        12  K = J
382            Y = W
383        13  CONTINUE
384        14  DO 15 J = 1,N
385            C(J) = A(J,K)
386            A(J,K) = A(J,I)
387            A(J,I) = - C(J)/Y
388            A(I,J) = A(I,J)/Y
389        15  B(J) = A(I,J)
390            A(I,I) = 1.0/Y
391            J = LZ(I)
392            LZ(I) = LZ(K)
393            LZ(K) = J
394            DO 19 K = 1,N
395            IF(I-K)16,19,16
396        16  DO 18 J = 1,N
397            IF(I-J)17,18,17
398        17  A(K,J) = A(K,J) - B(J)*C(K)
399        18  CONTINUE
400        19  CONTINUE
401        20  CONTINUE
402            DO 200 I = 1,N
403            IF(I-LZ(I))80,200,80
404        80  K = I + 1
405            DO 500 J = K,N
406            IF(I-LZ(J))500,600,500
407       600  M = LZ(I)
408            LZ(I) = LZ(J)
409            LZ(J) = M
410            DO 700 L = 1,N
411            C(L) = A(I,L)
412            A(I,L) = A(J,L)
413       700  A(J,L) = C(L)
414       500  CONTINUE
415       200  CONTINUE
416            RETURN
417            END
```

ÉPRT LANTA.INPUT

5

BINUCLEAR LANTHANIDE(III)–SILVER(I) NMR SHIFT REAGENTS

Thomas J. Wenzel

Bates College,
Department of Chemistry,
Lewiston, Maine 04240

Introduction

Lanthanide *tris-β*-diketonates were introduced as NMR shift reagents by Hinckley in 1969.[1] The lanthanide ion in these complexes is a hard Lewis acid and therefore bonds effectively to hard Lewis bases. Extensive reports followed in which the lanthanide *tris*-chelates were applied to donor compounds with hard Lewis base functional groups.[2–8] Soft Lewis bases, such as olefins, aromatics, halogenated compounds, and phosphines, do not bond effectively to the lanthanide ion in a *tris-β*-diketonate complex and the *tris*-chelates are not suitable shift reagents for these classes of compounds.

One approach to the successful development of NMR shift reagents for soft Lewis bases was demonstrated by Evans and co-workers.[9] By utilizing silver(I) heptafluorobutyrate in solution with lanthanide *tris-β*-diketonates they were able to observe shifts in the NMR spectra of olefins. A binuclear lanthanide(III)–silver(I) complex was formed in which the silver bonded to the soft Lewis base and the magnetic properties of the lanthanide caused the shifts in the NMR spectrum of the donor. In such

Morrill (ed): LANTHANIDE SHIFT REAGENTS IN STEREOCHEMICAL ANALYSIS

a binuclear complex the silver acts as a bridge between the lanthanide ion and the soft Lewis base. The donor is then held in some specific configuration relative to the lanthanide. The shifts observed for olefins in the presence of this binuclear complex are generally quite small and not of much practical utility for a broad range of NMR shift reagent applications.

Some workers did apply Evans' principle of a binuclear complex by employing silver(I)trifluoroacetate with lanthanide *tris*-β-diketonates. Meyers and Ford[10] were able to resolve partially the NMR spectra of the *d,l* enantiomers of methylene(2-methyl)cyclohexane with silver(I) trifluoroacetate and Eu(facam)$_3$.* Dambska and Janowski[11,12] have employed silver(I) trifluoroacetate with Pr(fod)$_3$ for the separation of the resonances of the isomers of the xylenes.

Binuclear lanthanide(III)–silver(I) complexes in which the silver species was either Ag(fod) or Ag(tfa) have been found to function more effectively as NMR shift reagents for soft Lewis bases than previously reported examples.[13] A number of followup reports extending the use and number of binuclear shift reagents involving a lanthanide *tris*-β-diketonate and silver β-diketonate have since appeared in the literature.[14–27] The application of these shift reagents to the study of olefins, aromatics, halogenated compounds, and phosphines has been demonstrated. Chiral analogs have been utilized successfully in the resolution of the NMR spectra of mixtures of +, − enantiomers. Table 5-1 gives the structure, name, and shorthand abbreviation for the ligands referred to in this chapter. As a result of the interest in binuclear NMR shift reagents, the compounds Ag(fod) and Ag(tfa) are now available from commercial sources.

Structural Considerations

A fundamental understanding of the structure of the binuclear complexes is not available at this time. It also should be stated that the reasons for one particular binuclear complex being more effective than another as an NMR shift reagent for soft Lewis bases are not thoroughly understood. Although this does not limit the application of the binuclear reagents in the simplification of complex NMR spectra, it does hinder

* Note that acac, tfa, hfa, thd, fod, dfhd, tta, hfth, facam and hfbc abbreviations are defined on Table 5-1.

TABLE 5-1. Ligands[a]

Name	Structure	Abbreviation
2,4-Pentanedione	$CH_3\overset{O}{\overset{\|}{C}}CH_2\overset{O}{\overset{\|}{C}}CH_3$	H (acac)
1,1,1-Trifluoro-2,4-pentanedione	$CF_3\overset{O}{\overset{\|}{C}}CH_2\overset{O}{\overset{\|}{C}}CH_3$	H (tfa)
1,1,1,5,5,5-Hexafluoro-2,4-pentanedione	$CF_3\overset{O}{\overset{\|}{C}}CH_2\overset{O}{\overset{\|}{C}}CF_3$	H (hfa)
2,2,6,6-Tetramethyl-3,5-heptanedione	$(CH_3)_3C\overset{O}{\overset{\|}{C}}CH_2\overset{O}{\overset{\|}{C}}C(CH_3)_3$	H (thd)[b]
6,6,7,7,8,8,8-Heptafluoro-2,2-dimethyl-3,5-octanedione	$(CH_3)_3C\overset{O}{\overset{\|}{C}}CH_2\overset{O}{\overset{\|}{C}}CF_2CF_2CF_3$	H (fod)
1,1,1,5,5,6,6,7,7,7-Decafluoro-2,4-heptanedione	$CF_3\overset{O}{\overset{\|}{C}}CH_2\overset{O}{\overset{\|}{C}}CF_2CF_2CF_3$	H (dfhd)
4,4,4-Trifluoro-1-(2-thienyl)-1,3-butanedione	S (ring) $-\overset{O}{\overset{\|}{C}}\,CH_2\overset{O}{\overset{\|}{C}}\,CF_3$	H (tta)
4,4,5,5,6,6,6-Heptafluoro-1-(2-thienyl)-1,3-hexanedione	S (ring) $-\overset{O}{\overset{\|}{C}}\,CH_2\overset{O}{\overset{\|}{C}}\,CF_2CF_2CF_3$	H (hfth)
3-(Trifluoroacetyl)-*d*-camphor	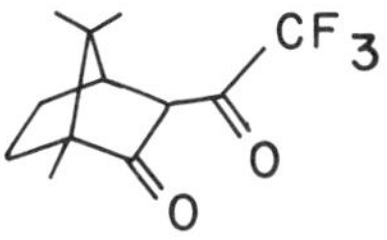 CF_3; O; O	H (facam)
3-(Heptafluoro-butyryl)-*d*-camphor	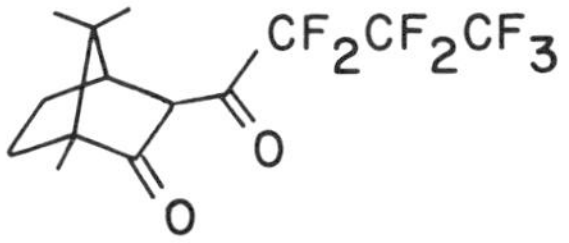 $CF_2CF_2CF_3$; O; O	H (hfbc)

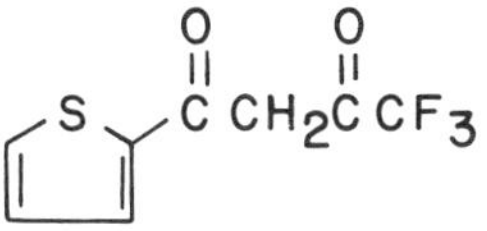

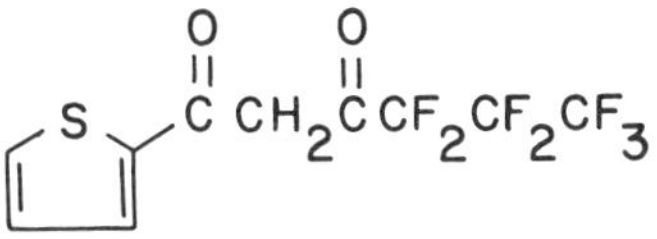

[a] The listed name is that of the parent ketone. The ligand corresponding to the parent ketone is named with an "onato" suffix replacing "one" (for example, acac is 2,4-pentanedionato). In all cases the ligand is the conjugate base of the parent ketone.
[b] This has been called H (dpm) by many authors.

their use in stereochemical studies in which the pseudocontact shift equation is to be rigorously applied.[28] These analyses require rather detailed information as to the distance from the lanthanide ion to the silver and olefin substrate, and the location of the principal magnetic axis of the shift reagent–silver–donor complex.

It appears that a 1:1 complex between the lanthanide *tris*-β-diketonate and silver β-diketonate is formed in solution.[15] In a solvent, such as $CDCl_3$, the solubility of the silver β-diketonate is enhanced considerably in the presence of the lanthanide *tris*-β-diketonate. The structure of the binuclear species may be similar to previously reported binuclear lanthanide–cesium complexes.[29–31] These complexes are ion pairs composed of a lanthanide tetrakis chelate anion and a cesium cation. The solid-state crystal structures of some of these have been reported in the literature.[29–32] If such a complex forms in solution with the binuclear shift reagents, the silver β-diketonate transfers its ligand to the lanthanide to form the tetrakis chelate anion, and the silver represents the cation of the ion pair.

It has been found that by taking a solution of $K[Eu(fod)_4]$ and toluene in $CDCl_3$ and shaking it with $AgBF_4$, an insoluble material, presumably KBF_4, separates out. The shifts observed in the resulting NMR spectrum of toluene in this solution are essentially identical to those observed with $Eu(fod)_3$–Ag(fod).[14]

If such an ion pair exists for the binuclear reagents, the mechanism by which the silver and lanthanide species attract one another is still uncertain. It could range from a purely electrostatic interaction of the silver cation with the lanthanide chelate anion, or could involve some type of bonding interaction. Possibilities include a situation in which two oxygen atoms of the β-diketonate ligands act as bridging atoms. This type of interaction has been observed for the binuclear lanthanide–cesium complexes.[29–31] The silver may bond to the π character of the chelate ring, a structure that has been observed for certain platinum complexes with β-diketone ligands.[33] The silver also may bond to the methine carbon of the β-diketone ligand, an interaction observed for binuclear nickel(II)–silver(I) β-diketonate complexes.[34] Until solid-state crystal structures of the binuclear lanthanide(III)–silver(I) complexes are determined their true natures must remain conjecture.

As mentioned at the beginning of this section, the reasons for the success of binuclear shift reagents involving only certain combinations of lanthanide *tris*-chelates and silver β-diketonates is not well understood. The improved solubility of many of the silver β-diketonates relative to the silver salts previously employed is certainly an important factor. It has been noted, however, that the magnitude of the induced shifts in the NMR spectrum of a substrate in the presence of the binuclear shift reagents is influenced by the choice of β-diketone ligands on both the lanthanide and the silver complex.[14] Combinations with similar solubilities result in considerably different shifts in the NMR spectrum of the substrate. As will be seen, this is especially evident in the selection of chiral binuclear shift reagents.

For applications in which the pseudocontact shift equation is to be applied rigorously, almost certainly a complex with four identical ligands would have to be used. The complications caused by a mixed ligand complex in the assumptions necessary for the pseudocontact shift equation are likely to present an intractable problem.

For applications in which only improved spectral dispersion is required, the best combinations of achiral shift reagents found to date are $Yb(fod)_3$ with Ag(fod) as a downfield shift reagent, and $Pr(fod)_3$ with either Ag(tfa) or Ag(tta) as an upfield shift reagent.[14,24] The shifts in the NMR spectrum of the substrate have always been found to be larger when $Yb(fod)_3$ was paired with Ag(fod) rather than Ag(tfa). With $Pr(fod)_3$, just the opposite is observed. The solubility of both silver β-diketonates with $Pr(fod)_3$ and $Yb(fod)_3$ is essentially identical.

One possible explanation for the observation that a change of only the ligand on the silver species leads to such a significant change in the induced shifts is that the association constant between the silver and the olefin may vary as the ligands vary. This mechanism has been largely ruled out, however, by measuring the shifts in the NMR spectrum of 1-hexene with the combinations $La(fod)_3$–Ag(fod), $La(fod)_3$–Ag(tfa), $Lu(fod)_3$–Ag(fod), and $Lu(fod)_3$–Ag(tfa) in $CDCl_3$.[14] Both lanthanum and lutetium were employed because the size of lanthanum is close to that of praseodymium and the size of lutetium is close to that of ytterbium. These combinations are all diamagnetic; therefore, any shifts observed in the NMR spectrum of 1-hexene result from complexation of the silver to the olefin. If the association constant changes with changes to the ligands or size of the lanthanide, it should be reflected in the magnitude of the complexation shifts. For all four cases, the shifts recorded in the NMR spectrum of 1-hexene were essentially identical, which indicated almost identical association constants. Of course these results only reflect association of the olefin to the silver and say nothing of the association between the silver–olefin species and the lanthanide.

A second explanation involves a change in the geometry of the shift reagent–substrate complex as the ligands are varied. Such a geometric change would be accounted for in the distance and angle terms in the pseudocontact shift equation. A ligand-dependent alteration of the geometry would change the induced shifts in the NMR spectrum of the substrate. It is also conceivable for the actual geometry of the shift reagent–substrate complex to remain identical from one binuclear complex to the next, but for the location of the principal magnetic axis, which is used in determining the angle term, to change from complex to complex as the ligands are varied. A third possibility is that the magnetic susceptibility of the lanthanide ion, another term incorporated into the pseudocontact shift equation, also is dependent on the nature of the ligands that make up the binuclear complex.

Lanthanide chelates with the ligands H(thd), H(acac), H(tfa), and H(hfa) were all found to be ineffective in binuclear applications with silver-β-diketonates.[18,21] The combinations of $Yb(fod)_3$ with Ag(hfa), Ag(thd), Ag(tta), Ag(hfth), or Ag(facam) and of $Pr(fod)_3$ with Ag(hfa), Ag(thd), Ag(hfth), or Ag(facam) were not as effective as the examples already mentioned.[18,21,24] For all of these combinations, the solubilities were roughly comparable.

Applications

Olefins

The effectiveness of the binuclear reagents has been demonstrated for a variety of applications involving olefins.[14,17,18,20,23–27] Binuclear shift reagents have been used to shift the ^1H NMR spectra of such olefins as 1-hexene, 4-vinyl-1-cyclohexene, 2,5-bicycloheptadiene, 1,6-diphenylhexatriene, 3,7-dimethylenebicyclo[3.3.1]nonane, α- and β-pinene, *d*-limonene, δ-3-carene, and camphene. They have been employed in the study of the ^{13}C NMR spectra of olefins.[17] Mixtures of cis and trans isomers also can be distinguished readily from each other in the presence of these reagents.

The NMR spectra obtained in a study of a mixture of *cis*- and *trans*-2-octene in the presence of the binuclear shift reagent $Yb(fod)_3$–Ag(fod) are shown in Figure 5-1. No apparent resolution of the two isomers is observed in the spectrum of *cis*- and *trans*-2-octene (0.1 M) in $CDCl_3$ without any shift reagent (Figure 5-1a). Addition of the binuclear reagent leads to resolution of the resonances of the cis and trans isomers; however, some unexpected behavior is observed. At a shift reagent concentration of 0.1 M (Figure 5-1b) the resonances for the cis isomer exhibit the larger shifts. Identification is based on the fact that the mixture is enriched in the cis isomer. The larger shifts for the cis isomer result from the competi-

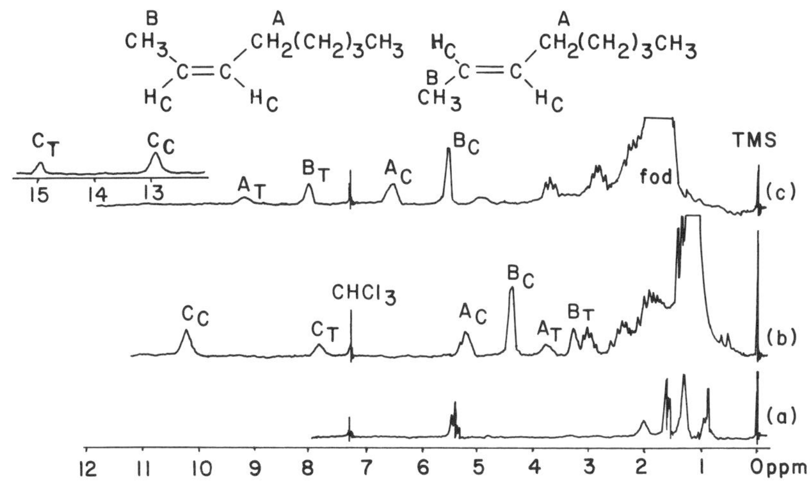

Figure 5-1. Proton NMR spectrum of 0.1 M *cis*- and *trans*-2-octene in $CDCl_3$ with (a) no shift reagent, (b) 0.1 M $Yb(fod)_3$ and 0.1 M Ag(fod), and (c) 0.2 M $Yb(fod)_3$ and 0.2 M Ag(fod).

tive effects of the two isomers for coordination sites on the shift reagent. The stability constant for association of the cis isomer with Ag(I) is known to be larger than that for the trans isomer.[35]

At a shift reagent concentration of 0.2 M (Figure 5-1c), the relative magnitude of the shifts of the two isomers is reversed: the larger resonances (cis isomer) now exhibit smaller shifts than those of the trans isomer. A similar result has been noted for *cis*- and *trans*-4-nonene and *cis*- and *trans*-2-hexene and therefore appears to be a general trend that is to be expected for all cis–trans olefin mixtures in the presence of the binuclear shift reagents.[14,27,36] The ability to separate the resonances of cis–trans isomers of olefins in the presence of the binuclear shift reagents has been applied to the analysis of hydrogenation products of alkynes.[25]

The reason for a reversal in the magnitude of the shifts for cis–trans isomers as the concentration of binuclear shift reagent is raised is not fully understood. At a concentration ratio with more shift reagent than substrate the competitive bonding effects must be removed. In the absence of competition it may be that the geometry (distance and angle terms) about the trans isomer result in larger shifts.

A more detailed study of this phenomenon has been reported for a mixture of *cis*- and *trans*-2-hexene.[27] In this study the binuclear relaxation reagent $Gd(fod)_3$–Ag(fod) was used to determine the location of the shift reagent relative to the substrate. The broadening of resonances in the presence of a relaxation reagent is dependent only on a distance term.[37] It was concluded from the relaxation and other data that the relative magnitudes of the induced shifts of *cis*- and *trans*-2-hexene in the abscence of competition cannot be explained by invoking only the distance term of the pseudocontact shift equation.

A shift mechanism other than the pseudocontact mechanism was postulated as a possible explanation for the results.[27] The internal ratio of the shifts for two different hydrogen atoms of the substrate *trans*-2-hexene in the presence of binuclear shift reagents containing europium, praseodymium, and ytterbium were calculated. This is a general procedure used to determine the presence of contact shifts. Contact shifts are indicated if the internal ratios vary as the metal is changed.[2,6,38] For *trans*-2-hexene a significant divergence in the ratios was observed as the metal of the shift reagent was changed. The ratio method assumes constant geometry for the shift reagent–substrate complexes for the entire series of lanthanides. Not enough is yet known about the structures of these complexes to determine whether this constraint is met. When the influence of the ligands on the effectiveness of the shift reagents is considered it seems unlikely that the geometries of all the $Ln(fod)_3$–Ag(fod)–*trans*-2-hexene complexes are the same. The differences in the induced shifts of *cis*- and *trans*-2-hexene not explained by distance considerations in fact may represent changes in the angle term of the pseudocontact equation.

For the binuclear reagents, application of the simplified form of the pseudocontact shift equation to the quantitative explanation of induced shifts may be entirely erroneous. If axial symmetry is not attained in these complexes, the more detailed form of the pseudocontact shift equation is necessary.

The terpenes provide an interesting class of compounds for study with these reagents.[18] They generally exhibit complicated NMR spectra and, because of their rigid ring systems, have a large number of nonequivalent hydrogen atoms. The shifts in the ^1H NMR spectra of α- and β-pinene, δ-3-carene, *d*-limonene, and camphene in the presence of $Yb(fod)_3$–Ag(fod) have been measured.[14,18] In all cases the shifted spectra were considerably more informative than the unshifted spectra. In these studies it was also noted that steric encumbrances of the ring system typically forced the silver to bond preferentially to one side of the olefin bond. A similar observation was reported by Smith[17] in a ^{13}C NMR study of α-pinene and norbornene. Smith's work also serves to demonstrate the suitability of applying the binuclear reagents to the study of the ^{13}C NMR spectra of olefins. In addition to α-pinene and norbornene, the ^{13}C NMR spectra of 2-methyl-2-butene, methylcyclohexene, and *syn*- and *anti*-sesquinorbornene in the presence of $Yb(fod)_3$–Ag(fod) were described.

Aromatics

The broadest demonstration of the applicability of binuclear NMR shift reagents to the study of soft Lewis bases has involved their use with aromatic substrates.[13,14,16,20–22,24] Substrates containing an aromatic moiety typically have NMR spectra that are not first order in nature. In many instances the NMR spectra of aromatic compounds appear deceptively simple. One example is the singlet observed for the aromatic hydrogen atoms in toluene. The bonding of aromatic compounds to Ag(I) is considerably weaker than that of olefins,[39] however, it is of sufficient strength to provide useful spectral clarification for a wide range of aromatic compounds in the presence of the binuclear reagents.

One rather simple example that demonstrates the power of the binuclear shift reagents is shown in Figure 5-2. Toluene has three non-equivalent aromatic hydrogen atoms. The resonances for the aromatic

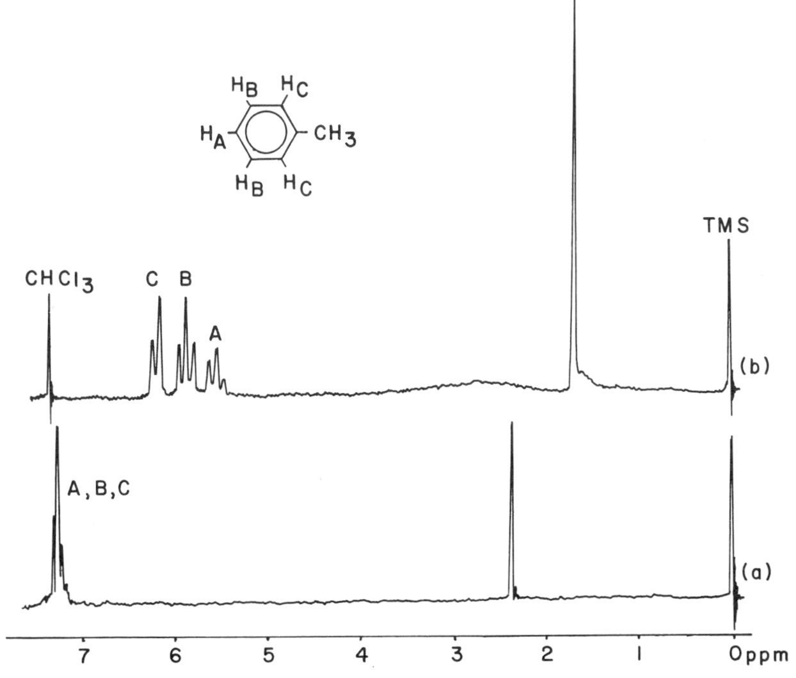

Figure 5-2. Proton NMR spectrum of 0.1 M toluene in $CDCl_3$ with (a) no shift reagent, and (b) 0.2 M $Pr(fod)_3$ and 0.2 M Ag(fod).

protons of toluene appear as one multiplet in a spectrum obtained at 90 MHz (Figure 5-2a). Addition of the binuclear shift reagent $Pr(fod)_3$–Ag(fod), a combination that produces upfield shifts in the NMR spectrum of a substrate, produces an NMR spectrum for toluene that is first order (Figure 5-2b). The relative magnitude of the shifts for the three nonequivalent aromatic protons reflects the preferential nature by which Ag(I) bonds to toluene. It is known that the bonding of Ag(I) to olefins and aromatics is highly influenced by the presence of steric encumbrances of the substrate.[39–41] For the compound toluene the Ag(I) prefers to bond at positions away from the methyl group. As a result the proton labeled A in the figure exhibits the largest shift, whereas proton C exhibits the smallest.

The coupling between the various protons also can be determined from the spectrum in Figure 5-2b. Proton A is a 1:2:1 triplet reflecting its coupling to the two B protons. Proton C is a 1:1 doublet because each of the protons labeled C couples to one B proton. The resonance for the two B protons are expected to be a doublet of doublets that result from the coupling of each to A and C. It appears as a 1:2:1 triplet because $J_{A\text{-}B}$ and $J_{B\text{-}C}$ are essentially identical. The resonances in the spectrum in Figure 5-2b do not show any evidence of long-range (four-bond) coupling. In fact, no NMR spectrum of aromatics recorded in the presence of binuclear reagents, such as $Pr(fod)_3$–Ag(fod) and $Yb(fod)_3$–Ag(fod), shows any evidence of long-range coupling. Four-bond coupling frequently is observed in the NMR spectra of aromatic compounds[42] and is probably absent from the shifted spectra because of the slight broadening that results in the resonances from the presence of the shift reagents.

The NMR spectra of methylbenzene derivatives, such as *o*-xylene, *m*-xylene, and 1,2,3-trimethylbenzene, are also more informative in the presence of the binuclear shift reagents.[21] The binuclear shift reagents can be utilized in the analysis of mixtures of methylbenzene derivatives. The methyl resonances in the NMR spectrum of a mixture of the three xylene isomers can be distinguished from each other in the presence of the binuclear shift reagents. Figure 5-3 shows an example in which $Pr(fod)_3$–Ag(fod) was added to an equimolar mixture of *o*-, *m*-, and *p*-xylene. Identification was achieved by selectively enriching the mixture with the three isomers. Mixtures containing methylbenzene and the three xylenes also can be quantitated in the presence of these shift reagents.[21]

The analysis of a more complicated mixture of methylbenzene derivatives is the study of unleaded gasoline shown in Figure 5-4. In the presence of $Yb(fod)_3$–Ag(fod) (Figure 5-4b) the methyl resonances for the

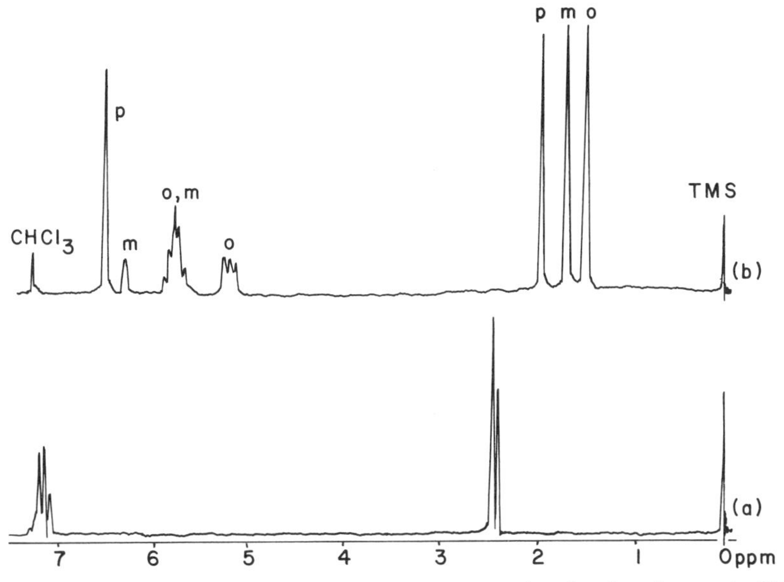

Figure 5-3. Proton NMR spectrum of a mixture 0.1 M of each xylene isomer in $CDCl_3$ with (a) no shift reagent, and (b) 0.3 M $Pr(fod)_3$ and 0.3 M Ag(fod).

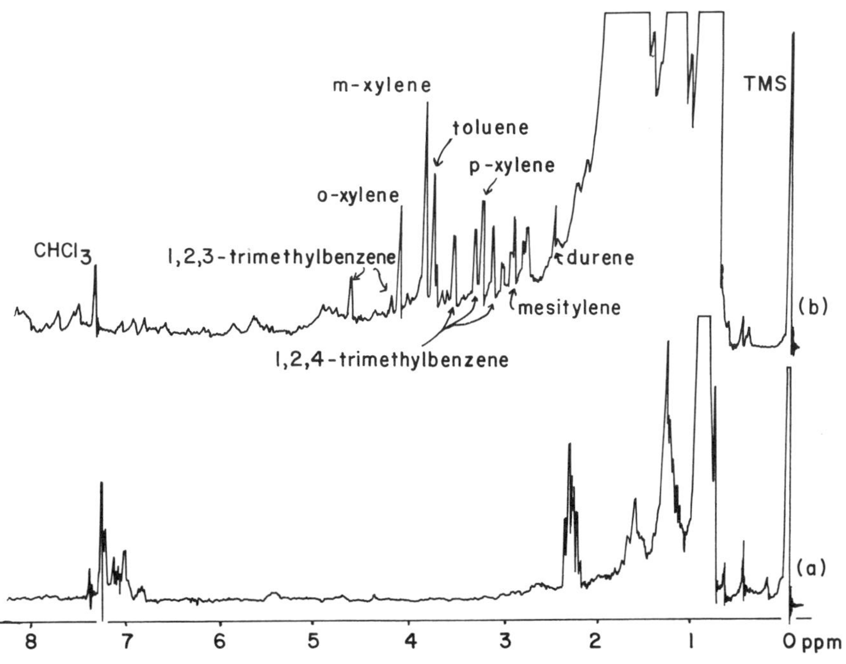

Figure 5-4. Proton NMR spectrum of 60 μL unleaded gasoline in 1 mL $CDCL_3$ with (a) no shift reagent, and (b) 0.3 M $Yb(fod)_3$ and 0.3 M Ag(fod). Reproduced with permission.[21]

compounds toluene, *o*-, *m*-, and *p*-xylene, 1,3,5-trimethylbenzene(mesitylene), 1,2,4-trimethylbenzene, and 1,2,4,5-tetramethylbenzene(durene) in unleaded gasoline have been distinguished and identified.

The binuclear shift reagents also have been applied to more complicated aromatic compounds, including *p*-terphenyl[13] and a series of fused ring polycyclic aromatics.[21] Polycyclic compounds that have been studied are naphthalene, 1-methylnaphthalene, acenaphthene, anthracene, 2-methylanthracene, 9-methylanthracene, phenanthrene, 1-methylphenanthrene, pyrene, benzo[*a*]pyrene, and 1,2:5,6-dibenzanthracene. In these studies it generally has been observed that only those compounds with one or more substituent groups on the ring system exhibit useful spectral clarification in the presence of the binuclear reagents. The NMR spectra of fused ring systems without a substituent group, naphthalene, anthracene, phenanthrene, pyrene, benzo[*a*]pyrene, and 1,2:5,6-dibenzanthracene, were shifted in the presence of the binuclear reagents; however, the bonding of the Ag(I) to the substrate was not selective enough in its location to provide especially selective shifts. Essentially the entire spectrum of these substrates in the presence of the binuclear reagents shifted as one unit without significant changes.

One example of the results observed for a polycyclic compound is shown in Figure 5-5 for the compound 9-methylanthracene. In this case a series of spectra was recorded as the concentration of $Yb(fod)_3$–Ag(fod) was increased from 0.1 to 0.5 M. As the series progresses it is possible to observe unique resonances at some point for each different aromatic proton in the substrate. This example demonstrates the value of performing an analysis in which the concentration of shift reagent is increased gradually relative to the substrate. In this series of spectra, there is no one example in which complete resolution is observed.

The binuclear shift reagents are also suitable for investigating the ^{13}C NMR spectra of aromatic compounds. The compound 1-methylnaphthalene has ten aromatic carbon atoms. In the ^{13}C NMR spectrum recorded at 100 MHz and 360 MHz only nine resonances are observed. In the presence of $Yb(fod)_3$–Ag(fod) (0.2 M) ten aromatic carbon resonances are observed in the ^{13}C NMR spectrum of 1-methylnaphthalene (0.1 M).[21]

The shifts in the ^{13}C NMR spectra of simpler aromatic compounds, such as toluene and *o*-xylene, have been also recorded.[21] In these spectra, which were obtained with the binuclear reagent $Eu(fod)_3$–Ag(fod), some of the ^{13}C resonances exhibited what would be considered wrong-way (upfield) shifts. The upfield shift results from complexation of Ag(I) to the aromatic.[43,44] The complexation shift can be measured by recording

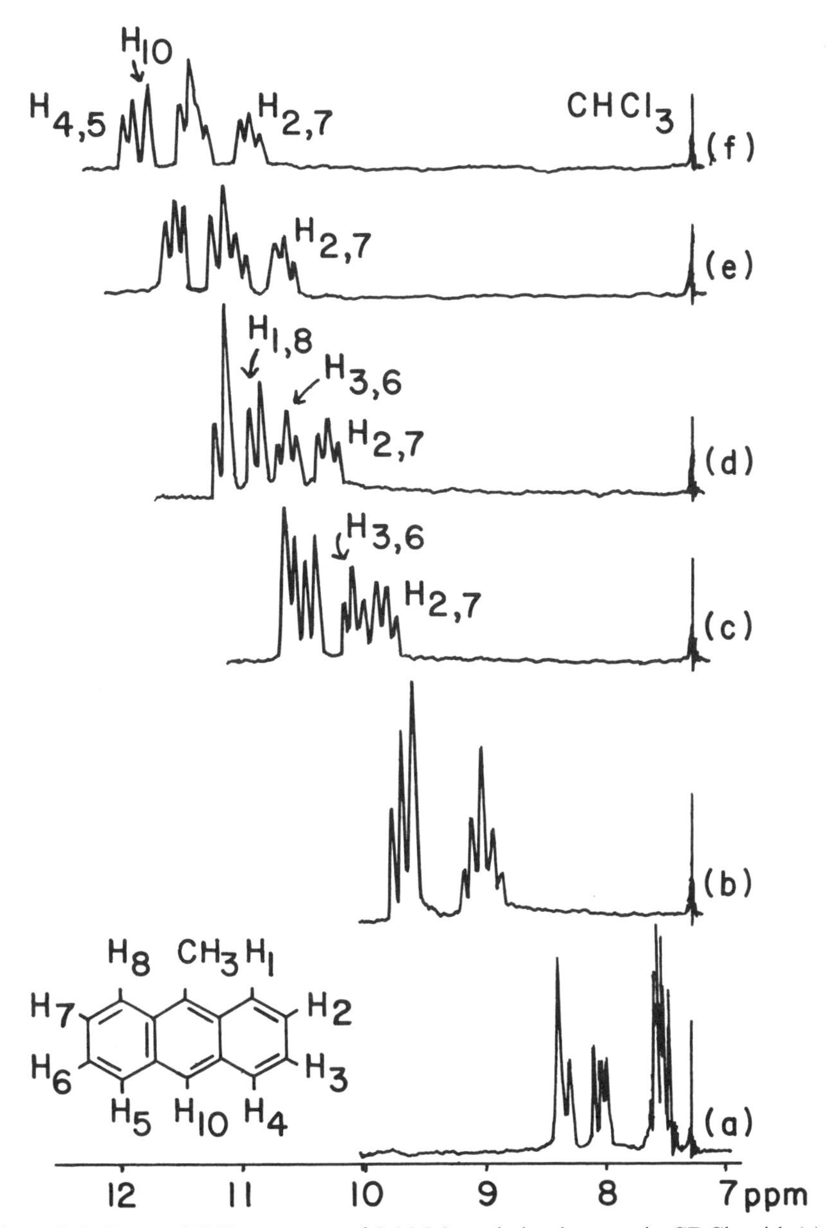

Figure 5-5. Proton NMR spectrum of 0.1 M 9-methylanthracene in $CDCl_3$ with (a) no shift reagent, (b) 0.1 M $Yb(fod)_3$ and 0.1 M Ag(fod), (c) 0.2 M $Yb(fod)_3$ and 0.2 M Ag(fod), (d) 0.3 M $Yb(fod)_3$ and 0.3 M Ag(fod), (e) 0.4 M $Yb(fod)_3$ and 0.4 M Ag(fod), and (e) 0.5 M $Yb(fod)_3$ and 0.5 M Ag(fod). Reproduced with permission.[21]

the ^{13}C NMR spectrum of the substrate in the presence of a diamagnetic binuclear complex such as $La(fod)_3$–Ag(fod). The value of the complexation shift may be important to assess when recording the ^{13}C NMR spectra of aromatic substrates in the presence of the binuclear reagents.

Halogenated Compounds

Most halogenated compounds bond only weakly to the lanthanide ion in a *tris*-β-diketonate complex. San Filippo has shown that the *tris*-chelates of fod and dfhd are useful NMR shift reagents for fluorinated compounds.[45] The *tris*-chelates are not effective NMR shift reagents, however, for chloro-, bromo-, and iodo-containing compounds. Halogenated compounds bond to Ag(I) and shifts can be recorded in the NMR spectra of certain chloro-, bromo-, and iodo-containing compounds in the presence of the binuclear reagents.[14] There are some limitations, however, on the general application of the binuclear reagents to organic halides that deserve mention.

In all cases studied the binuclear reagents have been found unsuitable for the study of secondary bromides and iodides, as well as tertiary chlorides, bromides, and iodides.[14] A precipitate, presumably the silver halide, forms on mixing these organic halides with the binuclear reagent. A second limitation is that the shifts observed in the NMR spectra of organic halides in the presence of the binuclear reagents are not large enough to be of practical utility unless the binuclear complex $Dy(fod)_3$–Ag(fod) is employed.[14] Dysprosium generally induces upfield shifts in the NMR spectrum of a substrate. It is the most powerful of all the lanthanide shift reagents and therefore causes considerable broadening of the resonances. For the organic halides, the induced broadening eliminates the fine structure caused by coupling in the spectrum.

The reason for only small shifts in the NMR spectra of organic halides in the presence of the binuclear shift reagents is not completely known but may be the result of a variety of factors. One possibility is that the association constant between the halide substrate and the silver is very small. This may be the case for chlorinated compounds, but is unlikely for the brominated and iodinated compounds. The compound iodobenzene is known to bond to silver at the iodine atom rather than the double bonds of the aromatic ring.[14,46] This indicates that iodo-containing compounds bond quite effectively to Ag(I), yet $Dy(fod)_3$–Ag(fod) is necessary to achieve useful shifts in the NMR spectra of iodinated compounds.

A second possibility is that the relatively large size of the halogen atom increases the distance between the lanthanide and the substrate necessitating a more powerful shift reagent. As Scheme I shows, the carbon and hydrogen atoms in the organic halide are further from the lanthanide than the compound with a carbon–carbon double bond. This same reasoning has been used to explain why europium is generally not effective enough in binuclear applications and why higher concentrations of the binuclear reagents are necessary to achieve useful spectral clarification than are needed in studies using only the tris chelates.[14,18]

Scheme I

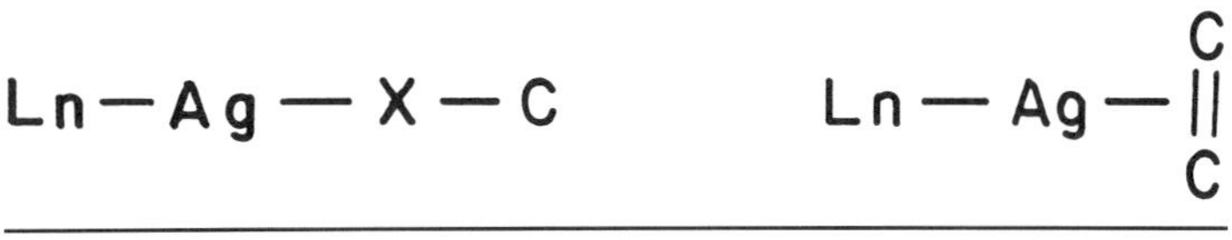

Ln—Ag—X—C   Ln—Ag—(C=C)

The utility of the binuclear shift reagents for organic halides has been demonstrated with the compounds chlorocyclohexane, chloropentane, bromopentane, and iodohexane.[14] For a series of compounds, such as the halobutanes, it has been found that the shifts are greatest for iodobutane and smallest for chlorobutane.[14] This results because the iodine atom is the softest Lewis base of the three and therefore has the largest association constant with Ag(I).

The NMR spectra obtained in a study of 1-iodohexane (0.1 M) in $CDCl_3$ are shown in Figure 5-6. The unshifted spectrum in Figure 5-6a is not first order because the resonances for the protons labeled B, C, and D overlap. In Figure 5-6b the spectrum recorded after addition of

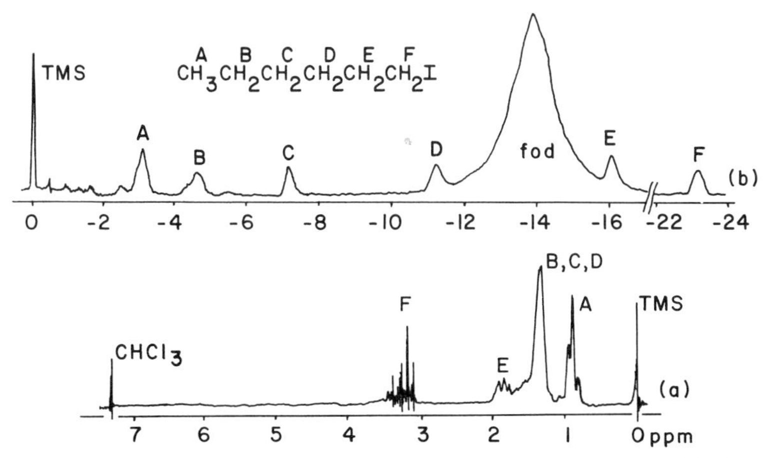

Figure 5-6. Proton NMR spectrum of 0.1 M 1-iodohexane in $CDCl_3$ with (a) no shift reagent, and (b) 0.2 M $Dy(fod)_3$ and 0.2 M Ag(fod).

the binuclear reagent $Dy(fod)_3$–Ag(fod) (0.2 M) is shown. In this spectrum the resonance for TMS is now furthest downfield. The protons labeled A, which are furthest downfield in the unshifted spectrum, are now the furthest upfield. A first-order spectrum is obtained with $Dy(fod)_3$–Ag(fod); however, the fine structure caused by coupling is absent from the spectrum. In instances in which the *t*-butyl resonance of the shift reagent interferes with the substrate resonances, $Dy(fod)_3$ can be replaced with $Dy(dfhd)_3$. The dfhd ligand is a more fully fluorinated ligand. In the $Dy(dfhd)_3$–Ag(fod) combination, the *t*-butyl resonance from the one-fod ligand appears downfield of TMS. As a result the spectrum obtained for an organohalide in the presence of $Dy(dfhd)_3$–Ag(fod) has no interfering resonances from the shift reagent.

Phosphines

Phosphines are soft Lewis bases that bond effectively to Ag(I).[47,48] Only very small shifts have been observed in the NMR spectra of phosphines in the presence of lanthanide *tris*-*β*-diketonates.[49,50] No noticeable shifts were observed in the NMR spectrum of triphenylphosphine (PPh_3) when lanthanide *tris*-*β*-diketonates are added.[51] In the presence of the binuclear shift reagent $Pr(fod)_3$–Ag(tfa), however, a first-order ^1H NMR spectrum is obtained for PPh_3.[21]

The spectrum observed for PPh_3 in the presence of the binuclear reagents is indicative of the silver preferentially bonding to the phosphorus atom rather than to the aromatic rings. Analysis of the ^{31}P spectrum confirms this. At 22°C with 0.2 M $Pr(fod)_3$, 0.2 M Ag(fod), and 0.15 M PPh_3 the phosphorus resonance appears as two doublets. These are the result of ^{107}Ag and ^{109}Ag coupling to the phosphorus. A similar coupling is observed in the ^{31}P NMR spectrum of triphenylphosphite when $Pr(fod)_3$–Ag(fod) is added. As the concentration of PPh_3 is increased an exchange of the PPh_3 that becomes rapid on the NMR time scale occurs and the resonance collapses to a singlet.

Chiral Substrates

By employing chiral lanthanide *tris*-*β*-diketonates it is possible to form chiral binuclear complexes capable of resolving the dextro and levo resonances of aromatic[19] and olefinic[10,18,19,23,24] enantiomers. The selection of the best chiral binuclear shift reagent for a particular substrate is a difficult process that is complicated because of the necessity of selecting both a lanthanide and a silver species. Studies have shown that the degree of enantiomeric resolution varies considerably with the nature of the silver species.[18,23,24]

There are two mechanisms that can cause enantiomeric resolution in the NMR spectrum of chiral donors in the presence of chiral shift reagents. The first is that the equilibrium constant for association between the dextro and levo enantiomer and the shift reagent can be different. The second is that the geometry of the complex between the dextro and levo enantiomer and the shift reagent can be different. Different geometries then can bring about different distance and angle terms in the pseudocontact shift equation. Either one or both mechanisms have been shown to be important for nitrogen- and oxygen-containing substrates in the presence of the *tris*-chelates.[52] It is reasonable to assume that both mechanisms are important in the enantiomeric resolution observed in the NMR spectra of soft Lewis bases in the presence of the binuclear shift reagents.

When performing chiral shift reagent studies it is best to record a series of spectra in which the concentration of the shift reagent is increased gradually relative to that of the substrate. This makes it possible to better distinguish enantiomeric resolution from the resolution of different protons of the substrate. When chiral studies are performed with either *tris*-β-diketonates or binuclear complexes, the enantiomeric resolution generally is observed for the resonances of protons situated close to the donor site of the molecule.[52] Also, because the extent of enantiomeric resolution is usually not that great, it is better to focus on the resonances of protons that are not coupled to many other protons.

The lanthanide *tris*-β-diketonates that have been used successfully to date in the formation of chiral binuclear reagents are those of the ligands H(facam) and H(hfbc). A rather wide variety of silver species has been employed in conjunction with these.

Meyers and Ford[10] have reported the partial resolution of the olefinic proton resonances in the compound methylene-(2-methyl)cyclohexane using $Eu(facam)_3$ with silver trifluoroacetate. Offermann and Mannschreck employed the europium, praseodymium, and ytterbium complexes of hfbc with Ag(fod) to study the degree of enantiomeric resolution in the ^1H NMR spectrum of α-pinene, camphene, limonene, and 3,4,5,6-tetramethylphenanthrene, and in the ^{13}C NMR spectra of α-pinene and camphene.[19] They reported better results for the ytterbium complex because of the general observation that ytterbium induces larger downfield shifts than europium. Krasutsky et al[23] reported slight enantiomeric resolution of some narrow peaks in the ^1H NMR spectrum of 3-methylene-7-benzylidenebicyclo[3.3.1]nonane. In this case,

$Eu(hfbc)_3$ was used with Ag(fod), Ag(hfbc), and $AgNO_3$ and only the binuclear complex with $AgNO_3$ resulted in enantiomeric resolution in the spectrum. The authors believe that this is the result of a significant difference in the bonding of the substrate to the silver (bidentate with $AgNO_3$, monodentate with the silver-β-diketonates) in the binuclear complexes.

The enantiomeric resolution in the ^1H NMR spectrum of the compounds α-pinene and camphene has been recorded using $Pr(facam)_3$, $Pr(hfbc)_3$, and $Yb(facam)_3$ with Ag(fod), Ag(tfa), and Ag(facam).[18] Although in most instances enantiomeric resolution was observed for one or more resonances, both the degree of enantiomeric resolution for a particular resonance and the resonances that were enantiomerically resolved varied with the nature of the silver species. This was especially noted with binuclear complexes formed with $Pr(hfbc)_3$.

Two particular examples worth noting are the results obtained for ($\pm$)-α-pinene (**1**) (0.1 M) with $Pr(hfbc)_3$–Ag(facam) (0.1 M) and for ($\pm$)-camphene (**2**) (0.1 M) with $Pr(hfbc)_3$–Ag(tfa) (0.05 M).[18] With α-pinene a larger shift is observed for the levo resonance of H_3, whereas for methyl

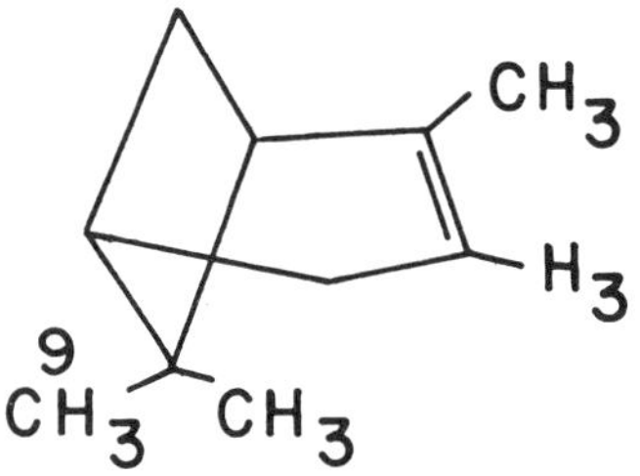

1 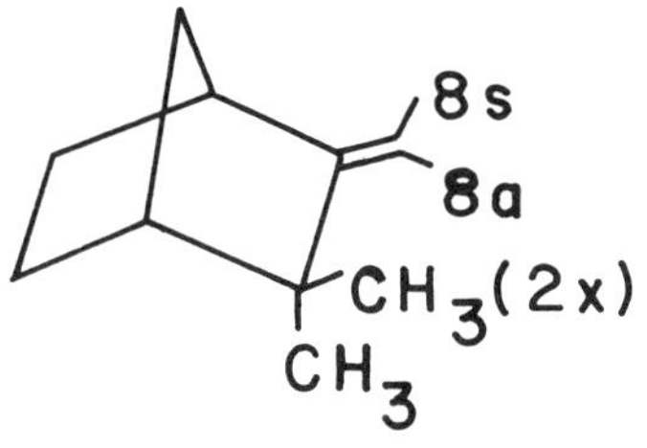 **2**

group 9 the dextro resonance exhibits the larger shift. For camphene the absolute configuration assigned to the resonances could not be determined, but a similar reversal in the magnitude of the shifts was observed when comparing the enantiomeric resolution of $2x$ relative to that of $8s$ and $8a$. This reversal of the relative magnitudes of the induced shifts of similar protons on the two enantiomers implies that for these two examples the enantiomeric resolution is the result of different geometries of the complexes between the shift reagent and the dextro and levo enantiomers. If the distinction were caused by differences in the association constants alone, all of the resonances of one enantiomer would be expected to exhibit the larger shifts.

The relative magnitudes of the induced shifts with chiral binuclear reagents involving $Pr(facam)_3$ is Ag(tfa) > Ag(fod) > Ag(facam). The degree of enantiomeric resolution, however, was not much different and was perhaps slightly better when Ag(facam) was employed.[18] For binuclear complexes with $Yb(facam)_3$, the results were quite similar when either Ag(fod) or Ag(tfa) was used. The combination of $Yb(facam)_3$ with Ag(facam), however, resulted in relative shifts and enantiomeric resolution clearly superior to that observed with any other chiral ytterbium binuclear reagent.[18]

The compounds Ag(tta) and Ag(hfth) have been tested with $Pr(facam)_3$ and $Pr(hfbc)_3$ for their ability to resolve the resonances of (±)-camphene.[24] In a comparison with the binuclear reagents formed with Ag(fod) and Ag(tfa) it was found that the combination of $Pr(facam)_3$ and $Pr(hfbc)_3$ with Ag(hfth) produced superior results. For binuclear reagents involving $Yb(facam)_3$, the results with Ag(tta) were slightly better than those with Ag(fod), Ag(tfa), and Ag(hfth), but not as good as those with Ag(facam).

Although the combination of $Yb(facam)_3$ with Ag(facam) results in the best enantiomeric resolution in the NMR spectrum of a substrate, it has some disadvantages that limit its use. The first is that it is not soluble at concentrations above 0.05 M. For many binuclear shift reagent applications it is desirable to have higher shift reagent concentrations than this. The second is that the compound Ag(facam) is relatively unstable when compared to most of the other silver β-diketonates. All of the silver β-diketonates appear to have a sensitivity toward light that is accentuated when in solution.[53,54] This does not hinder binuclear shift reagent studies, however, as it has been shown that solutions covered to exclude light can sit up to 4 days without any deleterious effects being observed.[14]

We have found Ag(facam) to exhibit a variable shelf life and to decompose on standing in a bottle that has been covered to exclude light. The compound Ag(hfbc) probably exhibits a similar instability. The one available report of Ag(hfbc) does not describe its synthesis,[23] and we have been unsuccessful in our laboratory in synthesizing and isolating this compound. The compounds Ag(tta) and Ag(hfth) are more stable in their solid state than the other silver β-diketonates. The increased stability of Ag(tta) and Ag(hfth) may result from an oligomerization in these compounds involving a silver bond to the sulfur atom on a neighboring molecule. A rough ordering of the relative stability is Ag(tta) > Ag(hfth) > Ag(fod) > Ag(tfa) > Ag(facam). All except Ag(facam) seem to have sufficient shelf lives for repetitive use. When Ag(facam) is to be employed, it is best to prepare it directly before use.

Based on the relative shifts, enantiomeric resolution, and stability, the first choice of an upfield chiral binuclear shift reagent at this time is

$Pr(hfbc)_3$ or $Pr(facam)_3$ with Ag(hfth). For a downfield shift reagent, the combinations of $Yb(facam)_3$ with Ag(tta) or of $Yb(hfbc)_3$ with Ag(fod) seem the best. It must be emphasized that no one chiral binuclear shift reagent necessarily functions most effectively with all substrates. Experimentation with a variety of binuclear complexes therefore may be necessary to achieve adequate enantiomeric resolution.

Polyfunctional Substrates

For compounds with two or more soft Lewis base donor groups, competition occurs between the donor groups for coordination sites on the shift reagent. A comparison of classes of soft Lewis bases indicates that olefins bond more effectively to Ag(I) than aromatics.[39] This is observed when the NMR spectra of the compounds styrene,[27] 4-methylstyrene,[9,14] and indene[27] are recorded in the presence of a binuclear shift reagent. In each instance the shifts are indicative of preferential bonding of the shift reagent to the olefin group. For styrene, at shift reagent–substrate ratios higher than 1, evidence for increased bonding of the binuclear reagent to the aromatic moiety is observed.[27]

The factors that influence the ability of olefin groups to bond to silver (I) are the localization of the π electrons, strain effects, and steric effects. Greater localization of the π electrons, larger relief of strain, and less steric hindrance favor a stronger complex between the silver and the olefin.[40,41] Such factors have been used to justify the preferential bonding of one of the olefin groups of the substrates (+)-limonene,[9,14,27] *cis*- and *trans*-hexa-2,4-diene,[27] and 5-methylenenorborn-2-ene[27] to the binuclear shift reagents. Analysis of the ^{1}H and ^{13}C NMR spectra of 4-vinyl-1-cyclohexene in the presence of the binuclear shift reagent $Eu(fod)_3$–Ag(fod) indicates a significant degree of complexation of both olefin groups.[14]

The term "multifunctional substrates" is used in this chapter to denote compounds that have both a hard and a soft Lewis base donor group. In many instances it would be advantageous to probe a multifunctional substrate at its various donor sites through the use of both binuclear and lanthanide NMR shift reagents. The question then arises as to whether the binuclear lanthanide–silver shift reagents can be used effectively in the presence of such substrates. In other words, will the binuclear shift reagent bond exclusively at the soft Lewis base donor site of such a compound? The answer to this question, although it is somewhat substrate dependent, appears to be no. It should be pointed out that there have been reports in which the binuclear reagents have been applied successfully to multifunctional substrates in which the hard Lewis base donor does not bond to the lanthanide in the *tris*-chelates.[16,20,27]

Studies to date indicate that the shifts recorded in the NMR spectrum of a multifunctional substrate reflect a degree of bonding of the binuclear reagent to the soft Lewis base site (silver to substrate), as well as a degree of bonding of a shift reagent species to the hard Lewis base site (lanthanide to substrate).[23,26,27,36] The nature of the species that bonds to the hard Lewis base site, whether binuclear or only lanthanide containing, has not been determined at this time. Although the binuclear reagents do not provide a means to probe only the soft Lewis base sites in the presence of hard Lewis base donors, the information to be learned by employing the binuclear reagents with such compounds may prove to be invaluable. Such studies, in the event that a *tris*-chelate does not yield the desired results, should be encouraged.

It has been shown that the binuclear reagents are useful for heteroaryl systems that do not exhibit shifts in the presence of lanthanide shift reagents[16,20,27]. These include indole, *N*-methylindole, pyrrole, benzothiophene, ethylbenzothiophene, benzofuran, methylbenzofurans, anisole, *p*-chlorophenylbenzoxazole, and phenylbenzoxazole. Evidence indicates that the silver of the binuclear reagent bonds to the sulfur atom in the benzothiophenes.[27] Binuclear reagents have also been used to probe cis-trans isomerism in the compound 2,2′-cyanine,[22] although no details of this study are presented in the report.

The shifts in the ^1H NMR spectrum of 1-hydroxymethyl-3,7-dimethylenebicyclo[3.3.1]nonane were recorded with the binuclear reagents $Eu(fod)_3$–$AgNO_3$ and $Eu(fod)_3$–Ag(fod).[23] The results with $Eu(fod)_3$–$AgNO_3$ were quite similar to those obtained with $Eu(fod)_3$. With Eu(fod)–Ag(fod) at shift reagent–substrate ratios less than 0.55, the magnitude of the shifts was indicative of the binuclear reagent bonding to the olefin group. At ratios higher than 0.55 the shifts reflect complexation of the europium at the hydroxy group.

The binuclear reagent $Eu(fod)_3$–Ag(fod) has been used to aid in the structural determination of the bitter principles extracted from mushrooms.[26] The natural products were converted to permethylethers prior to analysis with the binuclear reagent. Complexation of the shift reagent was indicated to occur at certain of the methoxy and olefin groups.

The compound linalool has also been studied with the binuclear reagent $Eu(fod)_3$–Ag(fod).[36] The observed shifts were indicative of complexation of a silver shift reagent species at the two olefin groups and a lanthanide species at the hydroxy group. The shifts observed in the NMR spectrum of 1,2-dimethoxybenzene in the presence of $Pr(fod)_3$–Ag(tfa) have been recorded.[27] The induced shifts were explained by a complexation of the binuclear species to the aromatic ring and a lanthanide species to the oxygen atoms. On increasing the shift reagent concentration, the induced shifts reflected increased bonding of the lanthanide species. This conclusion is analogous to that observed with 1-hydroxymethyl-3,7-dimethylenebicyclo[3.3.1]nonane.[23]

Experimental Aspects

The preparation and use of the binuclear shift reagents has been described in the literature.[14] The solvents benzene-d_6, acetone-d_6, acetonitrile-d_3, dimethyl-d_6-sulfoxide, and carbon disulfide are not suitable for use with the binuclear reagents. In the first four solvents only small shifts were observed for the substrate.[14] Addition of the silver species to CS_2 resulted in a chemical reaction. The solvents $CDCl_3$, CCl_4, and pentane are suitable for use with the binuclear reagents. It has been found recently that switching the solvent from $CDCl_3$ to pentane results in considerably larger shifts in the NMR spectra of chlorinated compounds (100%), aromatics (50%), brominated compounds (10–20%), and olefins (15%) in the presence of the binuclear reagents.[55] The pentane resonances overlap with the region from 1 to 4 ppm and cannot be used with subtrates that have shifted spectra with resonances in this region.

References

1. Hinckley, C. C. *J. Am. Chem. Soc.*, **1969**, *91*, 5160.
2. Sievers, R. E., Ed. "Nuclear Magnetic Resonance Shift Reagents"; Academic Press: New York, 1973.
3. Ċockerill, A. F.; Davies, G. L. O.; Harden, R. C.; Rackham, D. M. *Chem Rev.*, **1973**, *73*, 553.
4. Reuben, J. *Prog. Nucl. Magn. Reson. Spectrosc.*, **1975**, *9*, 1.
5. Flockhart, B. D. *CRC Crit. Rev. Anal. Chem.*, **1976**, *6*, 69.
6. Hofer, O. *Top. Stereochem.*, **1976**, *9*, 111.
7. Mayo, B. C. *Chem. Soc. Rev.*, **1973**, *2*, 49.
8. Inagaki, F.; Tatsuo, M.; *Prog. Nucl. Magn. Reson. Spectrosc.*, **1980**, *14*, 67.
9. Evans, D. F.; Tucker, J. N.; deVillardi, G. C. *J. Chem. Soc., Chem. Commun.*, **1975**, 205.
10. Meyers, A. I.; Ford, M. E. *J. Org. Chem.*, **1976**, *41*, 1735.
11. Dambska, A.; Janowski, A. *Org. Magn. Reson.*, **1980**, *13*, 122.
12. Dambska, A.; Janowski, A. *Chemia Analityczna*, **1980**, *25*,77.
13. Wenzel, T. J.; Bettes, T. C.; Sadlowski, J. E.; Sievers, R. E. *J. Am. Chem. Soc.*, **1980**, *102*, 5903.
14. Wenzel, T. J.; Sievers, R. E. *Anal. Chem.*, **1981**, *53*, 393.

15. Peters, J. A.; Schuyl, P. J. W.; Bovee, W. M. M. J.; Alberts, J. H.; van Bekkum, H. *J. Org. Chem.*, **1981**, *46*, 2784.
16. Rackham, D. M.; Crutchley, F. M.; Tupper, D. E.; Boddy, A. C. *Spectrosc. Lett.*, **1981**, *14*, 379.
17. Smith, W. B. *Org. Magn. Reson.*, **1981**, *17*, 124.
18. Wenzel, T. J.; Sievers, R. E. *J. Am. Chem. Soc.*, **1982**, *104*, 382.
19. Offermann, W.; Mannschreck, A. *Tetrahedron Lett.*, **1981**, *22*, 3227.
20. Rackham, D. M. *Spectrosc. Lett.*, **1981**, *14*, 639.
21. Wenzel, T. J.; Sievers, R. E. *Anal. Chem.*, **1982**, *54*, 1602.
22. Akins, D. L. *J. Coll. Int. Sci.*, **1982**, *90*, 373.
23. Krasutsky, P. A.; Yurchenko, A. G.; Radionov, V. N.; Jones, M., Jr. *Tetrahedron Lett.*, **1982**, *23*, 3719.
24. Wenzel, T. J.; Lalonde, D. R., Jr. *J. Org. Chem.*, **1983**, *48*, 1951.
25. McKenna, M.; Wright, L. L.; Miller, D. J.; Tanner, L.; Haltiwanger, R. C.; Dubois, M. R. *J. Am. Chem. Soc.*, **1983**, *105*, 5329.
26. Aoyagi, F.; Maeno, S.; Okuno, T.; Matsumoto, H.; Ikura, M.; Hikichi, K.; Matsumoto, T. *Tetrahedron Lett.*, **1983**, *24*, 1991.
27. Audit, M.; Demerseman, P.; Goasdoue, N.; Platzer, N. *Org. Magn. Reson.*, **1983**, *21*, 698.
28. McConnell, H. M.; Robertson, R. E. *J. Chem. Phys.*, **1958**, *29*, 1361.
29. Lippard, S. J.; Cotton, F. A.; Legdzins, P. *J. Am. Chem. Soc.*, **1966**, *88*, 5930.
30. Benette, M. J.; Cotton, F. A.; Legdzins, P.; Lippard, S. J. *Inorg. Chem.*, **1968**, *7*, 1770.
31. Burns, J. H.; Danford, M. D. *Inorg. Chem.*, **1969**, *8*, 1780.
32. Lalancette, R. A.; Cefola, M.; Hamilton, W. C.; La Placa, S. J. *Inorg. Chem.*, **1967**, *6*, 2127.
33. Allen, G.; Lewis, J.; Long, R. F.; Oldham, C. *Nature* (London), **1964**, *202*, 589.
34. Watson, W. H., Jr.; Lin, C. T. *Inorg. Chem.*, **1966**, *5*, 1074.
35. Hepner, F. R.; Trueblood, K. N.; Lucas, H. J. *J. Am. Chem. Soc.*, **1952**, *74*, 1333.
36. Wenzel, T. J. Ph. D. Thesis University of Colorado, Boulder, 1981.
37. Reuben, J.; Fiat, D. *J. Chem. Phys.*, **1969**, *51*, 4918.
38. Johnson, B. F. G.; Lewis, J.; McArdle, P.; Norton, J. R. *J. Chem. Soc., Chem. Commun.*, **1972**, 535.
39. Beverwijk, C. D. M.; van der Kerk, G. J. M.; Leusink, A. J.; Noltes, J. G. *Organomet. Chem. Rev., Sect. A*, **1970**, *5*, 215.
40. Muhs, M. A.; Weiss, F. T. *J. Am. Chem. Soc.*, **1962**, *84*, 4697.
41. Gil-Av, E.; Herling, J. *J. Phys. Chem.*, **1962**, *66*, 1208.
42. Silverstein, R. M.: Bassler, G. C.; Morrill, T. C. In "Spectrometric Identification of Organic Compounds," 3rd ed.; John Wiley and Sons: New York, 1974; p. 169.
43. van Dongen, J. P. C. M.; Beverwijk, C. D. M. *J. Organomet. Chem.*, **1973**, *51*, C36.
44. Christ, D. R.; Hsieh, Z. H.; Jordon, G. J.; Schinco, F. P.; Maciorowski, C. A. *J. Am. Chem. Soc.*, **1974**, *96*, 4932.
45. San Filippo, J., Jr.; Nuzzo, R. G.; Romano, L. J. *J. Am. Chem. Soc.*, **1975**, *97*, 2546.
46. Andrews, L. J.; Keefer, R. M. *J. Am. Chem. Soc.*, **1951**, *73*, 5733.
47. Gibson, D.; Johnson, B. F. G.; Lewis, J. *J. Chem. Soc. (A)*, **1970**, 367.
48. Partenheimer, W.; Johnson, E. H. *Inorg. Chem.*, **1973**, *12*, 1274.
49. Gerken, T. A.; Ritchey, W. M. *J. Magn. Reson.*, **1976**, *24*, 155.
50. Mandel, F. S.; Cox, R. H.; Taylor, R. C. *J. Magn. Reson.*, **1974**, *14*, 235.
51. Taylor, R. C.; Walters, D. B. *Tetrahedron Lett.*, **1972**, 63.
52. Sullivan, G. R. *Top. Stereochem.*, **1978**, *10*, 287, and references therein.
53. West, R.; Riley, R. *J. Inorg. Nucl. Chem.*, **1958**, *5*, 295.
54. Souza, S. M.; Nicholas, K. M. *Inorg. Chim. Acta*, **1979**, *33*, 77.
55. Wenzel, T. J. *J. Org. Chem.*, **1984**, *49*, 1834.

Author Index

Boldface numbers are reference number and chapter, respectively. Numbers in parentheses are pages on which cited.

A

Abraham, R. J., **29-4**, (111); **56-4**, (126) (128)
Ajisaka, K., **19-4**, (110)
Aki, O., **33-2**, (21)
Akins, D. L., **22-5**, (152) (156) (159) (171)
Alberts, J. H., **48-3**, (60); **15-5**, (154)
Alekel, R., **100-3**, (87)
Alexander, M., **60-4**, (129)
Allen G., **33-5**, (154)
Allinger, N. L., **45-4**, (119)
Ammon, H. L., **36-2**, (21)
Anderson, K. K., **32-2**, (21)
Andrews, L. J., **46-5**, (164)
Aoyagi, P., **26-5**, (152) (156) (171)
ApSimon, J. W., **71-2**, (31); **13-3**, (57); **116-3**, (93)
Archer, M. K., **33-3**, (57)
Arkko, A., **85-4**, (135) (137) (138)
Armitage,`I. M., **81-2**, (49); **12-3**, (57); **67-3**, (69) (71); **68-3**, (71); **101-3**, (87); **11-4**, (109) (129)
Aslanov, L. A., **26-3**, (57); **27-3**, (57)
Ateya, A., **31-2**, (21); **6-4**, (108) (119)
Audit, M., **27-5**, (152) (156) (157) (158) (170) (171) (172)
Ayräs, P., **28-4**, (110) (111) (112) (121)

B

Backus, J. J. M., **50-3**, (60); **59-3**, (61) (95)
Barciszewski, J., **5-2**, (49)
Barry, C. D., **33-1**, (6) (12); **42-1**, (12); **42-2**, (21)
Bassler, G. C., **42-5**, (160)
Bauer, S. H., **51-4**, (121)
Beauchamp, J. L., **104-3**, (89)
Beaumont, W. E., **44-3**, (58) (67) (68) (75) (77) (78)
Beaute, C., **37-1**, (8)
Behelfer, G. L., **37-1**, (8); **6-2**, (22); 77-3, (75)
Beierbeck, H., **71-2**, (31); **13-3**, (57); **116-3**, (93)
Bell, H. M., **58-4**, (128) (129)
Benette, M. J., **30-5**, (154)
Bernstein, H. J., **29-4**, (111)
Bettes, T. C., **3-5**, (151)
Beverwijk, C. D. M., **39-5**, (159) (160) (170)
Bilobran, D., **47-1**, (14)
Binks, J., **72-3**, (75) (78)
Birladeanu, I. **43-2**, (21)
Bleaney, B., 53-3, (61) (95); **3-4**, (107) (128)
Boddy, A. C., **16-5**, (152) (159) (171)
Boeyens, J. C. A., **52-1**, (15); **17-3**, (57); **22-3**, (57)
Bogel, M. E., **60-4**, (129)
Borg-Karlsson, A-K., **66-4**, (131); **67-4**, (131); **69-4**, (131)
Bouquant, J., **70-3**, (73) (91); **114-3**, (90); **115-3**, (91)
Bovee, W. M. M. J., **48-3**, (60); **49-3**, (60); **15-5**, (152) (154)
Boyd, W. A., **44-1**, (13)
Briggs, J. M., **15-2**, (21); **21-2**, (21); **25-2**, (21); **26-2**, (21); **30-2**, (21) (49); **37-2**, (21) (49); **59-2**, (24)
Brittain, H. G., **90-3**, **102-3**, (87); **103-3**, (87)
Brooks, J. J., **9-1**, (1); **49-1**, (15); **53-1**, (16); **23-2**, (21)
Brown, H. C., **21-4**, (110) (132) (135); **78-4**, (133); **79-4**, (133); **80-4**, (133); **86-4**, (133) (135); **87-4**, (133) (135); **88-4**, (133) (135)
Brown, J. M., **91-4**, (133) (135)
Bruder, A. H., **54-1**, (16); **87-3**
Brumley, W. C., **55-2**, (23)
Bundura, A. V., **51-2**, (22)
Burgett, C. A., **46-1**, (14)
Burkhard, J., **106-3**, (89)
Burns, J. H., 31-5, (154)
Burroughs, A. E., **60-4**, (129)

C

Caccamese, S., **47-2**, (22); **79-2**, (49)
Caines, G. H., **78-3**, (75)
Campbell, C. M., **38-3**, (58) (61) (67) (93); **42-3**, (58) (67) (68) (75) (77) (78) (87) (89) (96)
Campbell, J. R., **14-1**, (1)
Camps, P., **121-3**, (97)
Caple, R., **13-1**, (1); **69-2**, (26)
Carrea, G., **109-3**, (89)

Catton, G. A., **86-3**
Cawley, J. J., **62-3**, (64)
Cefola, M., **32-5**, (57)
Cerimele, B. J., **76-2**, (49)
Chadwick, D. J., **64-2**, (25); **65-2**, (25) (49)
Cheng, H. N., **57-3**, (61) (95)
Chiang, J. F., **51-4**, (121)
Christ, D. R., **44-5**, (162)
Chua, C., **60-4**, (129)
Chuche, J., **70-3**, (73) (91); **114-3**, (90); **115-3**, (91)
Clark, R. A., **47-1**, (14)
Clutter, D. R., **60-4**, (129)
Cockerill, A. F., **4-1**, (1); **57-2**, (24); **6-3**, (56) (58) (96); **71-3**, (89); **3-5**, (151)
Colicelli, E. J., **66-2**, (25)
Commeyras, **75-4**, (133); **77-4**, (133)
Cook, R. J., **65-4**, (131)
Cooper, R. A., **32-1**, (6); **46-2**, (21)
Coppell, S. M., **56-4**, (126)
Corey, E. J., **54-2**, (22); **46-4**, (119)
Corfield, P. W. R., **16-3**, (57)
Cotton, F. A., **19-3**, (57); **29-5**, (154); **30-5**, (154)
Couturier, J. E., **70-2**, (30) (49)
Cox, R. H., **50-5**, (166)
Cramer, R. E., **34-1**, (6); **24-3**, (57) (58); **41-3**, (58) (59); **46-3**, (59); **47-3**, (59)
Crane, R. W., **76-2**, (49)
Crippen, G. M., 52-2, (22)
Crump, D. R., **30-1**, (6); **118-3**, (96)
Crutchley, F. M., **16-5**, (152) (159) (171)
Cunningham, J. A., **9-1**, (1); **49-1**, (15); **53-1**, (16); **23-2**, (21); **15-3**, (57); **20-3**, (57); **40-3**, (58)

D

Dambska, A., **11-5**, (152); **12-5**, (152)
Danford, M. D., **31-5**, (154)
D'Angelo, A., **44-2**, (21); **76-3**, (75)
David, C. W., **104-4**, (140)
Davis, S., **117-3**, (96)
Davies, G. L. O., **57-2**, (24); **6-3**, (56) (58) (96); **1-4**, (107); **3-5**, (151)
Davis, R. E., **40-1**, (10); **9-2**, (31); **15-2**, (21); **43-2**, (21); **73-2**, (49); **123-3**, (97); **20-4**, (110); **55-4**, (126)
De Boer, E., **20-1**, (2); **24-2**, (21); **36-3**, (57) (96); **50-3**, (60); **59-3**, (61) (95)
De Boer, J. W. M., **24-2**, (21); **36-3**, (57) (96)
De Gennaro, N. K., **44-2**, (21); **76-3**, (75)
DeHorrocks, W., Jr., **45-1**, (13)
DeMarco, P. V., **25-1**, (5) (6); **32-2**, (21); **76-2**, (49); **64-3**, (64); **10-4**, (108) (115) (137)
DeMember, J. M., **77-4**, (133); **99-4**, (135)
Demerseman, P., **27-5**, (152) (156) (157) (158) (170) (171) (172)
Deranleau, D. A., **81-3**, (76) (77) (92)
DeSalas, **103-4**, (137)
Desreux, J. F., **34-3**, (57)
De Villiers, J. P. R., **52-1**, (15); **22-3**, (57)
DeVillardi, G. C., **3-5**, (151)
De Waard, E. R., **109-3**, (89) (96)
Dietrich, W., **63-4**, (129)
Dillen, J., **49-2**, (22)
Dobson, C. M., **33-1**, (6); **42-2**, (21)
Doering, W. von E., **43-2**, (21)
Dorofeeva, I. B., **7-2**, (21)
Drew, M. G. B., **51-3**, (60)
Dubey, R., **113-3**, (89)
Dubois, R., **41-3**, (58) (59); **46-3**, (59); **47-3**, (59); **25-5**, (152) (156) (157)
Duddeck, H., **63-4**, (129)
Dunsmore, G., **67-3**, (131)
Dyer, D. S., **49-1**, (15); **53-1**, (16); **23-2**, (21)

E

Early, T. A., **19-2**, (21)
Eisentraut, K. J., **27-1**, (6)
Elgavish, G. A., **7-3**, (56) (58); **58-3**, (61) (95); **82-3**, (76) (77) (92)
Elzey, T. K., **25-1**, (5) (6); **64-3**, (64); **10-4**, (108) (115) (137)
Emch, R., **60-4**, (129)
Erasmus, C. S., **17-3**, (57)
Ernst, L. **112-3**, (89)
Ernst, R. R. J., **61-2**, (24)
Evans, D. F., **22-1**, (3) (15); **30-3**, (57) (59); **45-3**, (59); **9-5**, (151) (170)
Everling, B. W., **104-4**, (140)
Eyring, H., **11-2**, (21)

F

Fabre, O., **92-3**
Farid, S., **31-2**, (21); **6-4**, (108) (119)
Feibush, B., **32-3**, (57)
Fell, D. S., **33-3**, (21)
Fiat, T., **1-4**, (107); **37-5**, (158)
Finocchiaro, P., **63-2**, (25); **77-2**, (49); **72-4**, (131)
Fischer, R. D., **7-1**, (1) (6) (10)
Fischer, W., **96-4**, (135)
Flockhart, B. D., **5-5**, (151)

Fomichey, A. F., **74-2**, (49)
Font, J., **121-3**, (97)
Ford, L. E., **33-1**, (6) (12); **42-2**, (21); **3-5**, (151)
Forsén, K., **16-4**, (105) (119) (123); **49-4**, (119)
Fox, L. E., **34-3**, (57)
Francis, H. E., **46-1**, (14)
Fruchier, A., **13-3**, (57); **116-3**, (93)
Fujiwara, H., **84-3**, (89); **89-3**
Furman, B., **48-1**, (14)

G

Gansow, O. A., **40-1**, (10); **20-4**, (110)
Geise, H. J., **49-2**, (22)
Gerber, N. N., **40-4**, (115)
Gerken, T. A., **49-5**, (166)
Gibb, V. G., **12-3**, (57); **101-3**, (87)
Gibson, D., **47-5**, (166)
Gil-Av, E., **41-5**, (160) (170)
Gilman, L., **84-2**, (50)
Girard, P., **117-3**, (96)
Glasel, J. A., **17-1**, (1); **42-1**, (12)
Glusker, J. P., **111-3**, (89)
Goasdoue, N., **27-5**, (152) (156) (157) (158) (170) (171) (172)
Godwin, A. D., **31-3**, (57) (65) (76) (90); **12-4**, (109) (110), **18-4**, (110) (129)
Golding, R. M., **55-3**, (61) (95); **2-4**, (107), **4-4**, (107), (128)
Graddon, D. P., **94-3**, 97-3
Gray, A. I. **68-2**, (25)
Green, M. M., **65-4**, (129)
Griengl, H., **119-3**, (97)
Grob, C. A., **95-4**, (135); **96-4**, (135); **97-4**, (135); **100-4**, (135)
Grosse, M., **9-3**, (57); **91-3**
Grotens, A. M., **50-3**, (60); **59-3**, (61) (95)
Groves, J. T., **64-4**, (129) (130)
Guida, A., **42-3**, (58) (67) (68) (75) (77) (78) (87) (89) (96)
Günther, B., **97-4**, (135)
Gutowski, H. S., **57-3**, (61) (95)

H

Hájek, M., **37-3**, (57); **106-3**, (89); **62-4**, (129)
Häkli, H., **7-4**, (108); **16-4**, (109) (119) (123); **17-4**, (109) (119) (126); **54-4**, (123) (125); **73-4**, (132) (133)
Haky, J. E., **94-4**, (135)
Hall, L. D., **81-2**, (49); **12-3**, (?): **67-3**, (69) (71); **68-3**, (69) (71); **101-3**, (87); **11-4**, (109) (129)
Halonen, A., **23-4**, (110)
Haltiwanger, R. C., **25-5**, (152) (157)
Halton, M. P., **2-4** (107)
Hamilton, W. C., 28-2, (21); **32-5**, (154)
Hammons, J. H., **92-4**, (135)
Hanreich, A., **96-4**, (135); **97-4**, (135)
Hanson, S. W., **29-1**, (5) (6); **10-3**, (57) (62) (64)
Hardee, L. E., **61-3**, (64) (65) (66) (67) (69) (70) (73) (74) (75) (76) (91) (92)
Harden, R. C., **4-1**, (1); **57-2**, (24); **6-3**, (56) (58) (96); **3-5**, (151)
Harris(s), D. H., **13-1**, (1); **69-2**, (26)
Hart, F. A., **21-2**, (21); **25-2**, (21); **59-2**, (24); **86-3**
Hart, H., **108-3**, (89)
Havel, T. F., **52-2**, (22)
Hawkes, G. E., **31-1**, (6); **32-1**, (6); **3-2**, (21) (49); **17-2**, (21) (24); **46-2**, (21); **72-2** (49)
Hayashi, S., **50-4**, (119)
Hazek, M., **58-2**, (24)
Heigl, T., **62-2**, (24); **78-2**, (49)
Helske, M., **31-4**, (112) (115)
Hepner, F. R., **35-5**, (157)
Herling, J., **41-5**, (160) (170)
Herwig, K., **32-1**, (6); **46-2**, (21)
Heubucek, J. R., **35-2**, (21)
Hilichi, K., **26-5**, (152) (156) (171)
Hilbers, C. W., **24-2**, (21); **36-3**, (57) (96)
Hilderbrandt, R. L., **12-2**, (21)
Hinckley, C. C., **1-1**, (1); **44-1**, (13); **55-2**, (23); **80-2**, (49); **1-3**, (55) (97); **1-5**, (151)
Hirayama, M., **95-3**, **96-3**
Hirsjärvi, P., **23-4**, (110); **24-4**, (110); **25-4**, (108)
Hlubucek, J. R., **36-1**, (8) (9)
Hoard, J. L., **28-3**, (57); **29-3**, (57)
Hofer, O., **6-1**, (1) (6); **3-3**, (56) (58); **99-3**, (87) (96); **119-3**, (97); **6-5**, (151) (158)
Holder, R. W., **55-4**, (126)
Honeyburne, C. L., **38-2**, (58) (61) (93)
Horrocks, W. deW., Jr., **35-1**, (6); **16-2**, (21); **29-2**, (21); **25-3**, (57) (58)
Hsieh, Z. H., **44-5**, (162)
Huber, H., 40-2, (21); 83-2, (49)
Hudson, M. J., **51-3**, (60)
Huet, J., **92-3**
Huisman, H. O., **109-3**, (89) (96)
Husar, J., **72-2**, (49)

I

Iilinskii, A. L., **26-3**, (57)
Ikura, M., **26-5**, (152) (156) (171)

Inagaki, F., **19-1**, (1) (11); **55-1**, (16); **2-3**, (56) (58); **79-3**, (75); **80-3**, (75); **8-5**, (151)
Ius, A., **109-3**, (89)
Ivanov, V. I., **26-3**, (57)

J

Jack, A., **4-2**, (21)
Jackson, G. F., III, **42-3**, (58) (67) (68) (87) (89) (96), **65-3**, (67) (75) (77) (89) (91); **74-3**, (75)
Janks, C. M., **6-2**, (22), **41-2**, (58); **38-3**, (58) (67) (93); **42-3**, (58) (67) (68) (75) (77) (78) (87) (89) (96); **65-3**, (65) (75) (77) (89) (91); **66-3**, (67) (75) (77) (91) (94); **77-3**, (75)
Jankowski, K., **2-2**, (22); **70-2**, (30) (49)
Janků, J., **106-3**, (89)
Janowski, K., **3-5**, (151)
Jantzen, R., **37-1**, (8)
Januszewski, H., **39-1**, (10) (12)
Jenkins, D., **41-4**, (115)
Johns, S. R., **32-1**, (6); **17-2**, (21) (24); **46-2**, (21)
Johnson, B. F. G., **38-5**, (158); **47-5**, (166)
Johnson, L. F., 36-1, (8) (9); **35-2**, (21); **48-5**, (166)
Johnson, M. D., Jr., **6-2**, (22); **41-2**, (21); **42-2**, (21); **44-2**, (21); **11-3**, (57) (75) (76); **31-3**, (57) (65) (76); **38-3**, (58) (61) (67) (93); **42-3**, (58) (67) (68) (75) (77) (78) (87) (89) (96); **43-3**, (58) (67) (68) (75) (77) (78); **44-3**, (58) (67) (68) (75) (77) (78); **65-3**, (67) (75) (77) (89) (91); **66-3**, (67) (75) (77) (91) (94); **74-3**, (75); **75-3**, (75); **76-3**, (75); **76-3**, (75); **77-3**, (75); **12-4**, (109) (110); **18-4**, (110) (129)
Jones, M., Jr., **2-3**, (56) (58)
Jordan, G. J., **44-5**, (162)
Jotham, R. W., **33-3**, (57)

K

Kadai, I., **90-4**, (135)
Kagan, H., **117-3**, (96)
Kainosho, M., **19-4**, (110)
Kainulainen, A., **23-4**, (110)
Karham, **37-3**, (57)
Karonski, M., **5-2**, (49)
Karraker, D. G., **18-3**, (57)
Kawakami, J. H., **80-4**, (133); **88-4**, (133) (135)
Kawaki, H., **84-3**, (89); **89-3**
Keefer, R. M., **46-5**, (164)
Kelsey, D. R., **69-3**, (69) (73)
Khalilov, **26-3**, (57)
Kieboom, A. P. G., **8-3**, (56) (58) (91)
Kilian, R. J., **104-4**, (140)
Kime, K. A., **15-1**, (1) (13) (15)
Kiss, A. I., **50-2**, (22)
Kleinfelter, D. C., **83-4**, (133)
Knihtila, H., **85-4**, (133) (137) (138)
Koch, C., **41-4**, (115)
Kopecky, W. J., Jr., **36-2**, (21) (25)
Koppel, I., **104-3**, (89)
Kordova, I., **74-2**, (49)
Korvola, J., **73-4**, (132) (133)
Korytnii, E. F., **27-3**, (57)
Kosobutskii, V. A., **7-2**, (21)
Krasutsky, P. A., **23-5**, (152) (156) (166) (167) (169) (171) (172)
Kruk, C., **109-3**, (89) (96)
Ksandr. Z., **58-2**, (24); **62-4**, (129)
Kuo, S. C., **13-1**, (1); **69-2**, (26)
Kuorikoski, S-L., **101-4**, (135) (137)

L

Laasonen, T., **101-4**, (135) (137)
Laatikainen, R., **52-4**, (121)
Lalancette, R. A., **32-5**, (154)
La Placa, S. J., **32-5**, (154)
Lalonde, D. R., Jr., **24-5**, (152) (155) (159) (166) (167) (169)
Lee, B., **28-3**, (57); **29-3**, (57); **120-3**, (97)
Legzdins, P., **19-3**, (57); **29-5**, (154); **30-5**, (154)
Leibfritz, D., **31-1**, (6); **32-1**, (6); **3-2**, (21); **46-2**, (21)
Lenkinski, R. E., **40-1**, (10); **18-2**, (21) (22) (24) (49); **43-2**, (21); **67-2**, (25); **82-3**, (76) (77) (92); **20-4**, (110); **71-4**, (131)
LePage, W. R., **85-2**, (50)
Leusink, A. J., **39-5**, (159) (170)
Levine, B. A., **43-1**, (12); **56-2**, (23)
Lewin, A. H., **100-3**, (87)
Lewis, J., **33-5**, (154); **38-5**, (158); **47-5**, (166)
Lewis, R. B., **25-1**, (5) (6); **64-3**, (64); **10-4**, (108) (115) (137); **61-4**, (129)
Li, W.-K., **27-2**, (21) (49)
Librando, V., **47-2**, (22); **79-2**, (49)
Lin, C. T., **34-5**, (154)
Lind, M. D., **28-3**, (57); **29-3**, (57)
Linder, M., **61-2**, (24)
Lindoy, L. F., **51-3**, (60); **52-3**, (60); **97-3**
Lip, H. C., **51-3**, (60)
Lippard, S. J., **29-5**, (154); **30-5**, (154)

Liu, K.-T., **80-4**, (133); **86-4**, (133) (135); **89-4**, (133) (135); **93-4**, (135)
Loeffler, P. A., **40-1**, (10); **20-4**, (110)
Long, R. F., **33-5**, (154)
Lopata, A., **50-2**, (22)
Lorentz, D. R., 16-1, (1)
Louie, H. W., **51-3**, (60); **52-3**, (60); **97-3**
Love, G. M., **108-3**, (89)
Luber, J. R., **45-1**, (13); **26-3**, (57)
Lucas, H. J., **35-5**
Lui, C. Y., 75-4, (133); **77-4**, (133); **99-4** (135)

M

Maciorowski, C. A., **44-5**, (162)
Maeno, S., **26-5**, (152) (156) (171)
Maggio, M., **6-4**, (108)
Mäkelä, R., **85-4**, (133) (135) (137)
Mälkönen, P. J., **30-4**, (112) (115) (125); **31-4**, (112) (115); **35-4**, (112) (115); **36-4**, (112) (115); **37-4**, (112) (115) (125); **38-4**, (115); **39-4**, (115)
Mandel, F. S., **50-5**, (166)
Mannschreck, A., **112-3**, (89); **19-5**, (152) (166) (167)
Maravigna, P., **63-2**, (25); **79-2**, (49)
Mariano, P. S., **59-4**, (128)
Marinetti, T. D., **20-2**, (21)
Marks, T. J., **35-3**, (57) (58)
Marques, J. M., **121-3**, (97)
Marshall, A. G., **81-2**, (49); **12-3**, (57); **67-3**, (69) (71); **68-3**, (69) (71); **101-3**, (87); **11-4**, (109) (129)
Marzin, C., **32-1**, (6); **17-2**, (21) (24); **46-2**, (21)
Maslo, M., **31-2**, (21)
Matsubara, Y., **50-4**, (119)
Matsumoto, H., **26-5**, (152) (171)
Matsumoto, T., **26-5**, (152) (171)
Matsuo, A., **50-4**, (119)
Maugean, A., **114-3**, (90)
Maynard, R. B., **41-3**, (58) (59)
Mayo, B. C., **3-1**, (1) (10); **5-3**, (56) (58); **7-5**, (151)
Mazzochi, P. H., **36-2**, (21) (25); **66-2**, (25)
McArdle, P., **120-3**, (97); **38-5**, (158)
McConnell, H. M., **1-2**, (21); **5-4**, (108), **28-5**, (154)
McElroy, R., **59-4**, (128)
McGarvey, B. R., **54-3**, (61) (95)
McIver, R. T., Jr., **104-3**, (89)
McIvor, M. C., **91-4**, (135)
McKenna, M., **25-5**, (152) (157)
Medsker, L. L., **41-4**, (115)
Meek, D. W., **28-1**, (6)
Meerwein, H., **102-4**, (137)
Melaja, A., **23-4**, (110)
Meravigana, P., **72-4**, (131)
Meyers, A. I., **10-5**, (152) (166) (167)
Micklow, G. K., **62-2**, (24)
Miller, D. J., **25-5**, (152) (157)
Miller, T. C., **42-5**, (160)
Miwa, M., **55-1**, (16)
Miyazawa, T., **19-1**, (1) (11); **35-1**, (6); **2-3**, (56) (58); **79-3**, (75); **80-3**, (75)
Mizuta, E., **33-2**, (21)
Mohyla, I., **58-2**, (24); **62-4**, (129)
Montando, G., **47-2**, (22); **63-2**, (25); **77-2**, (49); **79-2**, (49); **72-4**, (131)
Monti, C. T., **111-3**, (89)
Morallee, K. G., **75-2**, (49)
Morgan, L. O., **56-3**, (61) (95)
Morimoto, A., **33-2**, (21)
Morrill, T. C., **24-1**, (14); **47-1**, (14); **48-1**, (14); **105-3**, (89)
Moss, G. P., **15-2**, (21); **21-2**, (21); **25-2**, (21); **59-2**, (24); **86-3**
Mozzer, R., **24-1**, (14); **105-3**, (89)
Mucklow, G. C., **78-2**, (49)
Muhs, M. A., **40-5**, (160) (170)
Muir, L., **94-3**, **97-3**
Murray-Rust, J., **111-3**, (89)
Musher, J. L., **27-4**, (110) (112)

N

Nakayama, M., **50-4**, (119)
Nevell, T. P., **103-4**, (137)
Newman, R. H., **39-2**, (21)
Nicholas, K. M., **54-4**, (169)
Nickon, A., **92-4**, (135)
Nieboer, E., **75-2**, (49)
Nieuwenhuizen, M. S., **125-3**, (95)
Noltes, J. G., **39-5**, (160) (170)
Nordlander, J. E., **14-2**, (22); **94-4**, (135)
Norin, T., **66-4**, (131); **68-4**, (131); **69-4**, (131)
North, C. T., **42-1**, (12)
Norton, J. R., **38-5**, (158)
Novogelev, N. P., **51-2**, (22)
Nowak, P., **119-3**, (97)
Nuzzo, R. G., **110-3**, (89); **45-5**, (164)

O

Ochiai, M., 33-2, (21)
Offermann, W., **19-5**, (152) (166) (167)

Ojala, K., **31-4**, (112) (115); **32-4**, (112) (115)
Okada, T., **33-2**, (21)
Okuno, T., **26-5**, (152) (171)
Olah, G. A., **75-4**, (133); **76-4**, (133); **77-4**, (133); **99-4**, (135)
Oldham, C., **33-5**, (154)
Opitz, R. J., **24-1**, (14); **105-3**, (89)
O'Reilly, J. P., **120-3**, (97)
Orrell, K. G., **18-1**, (1) (10)
Owada, M., **95-3**
Owuoi, P. O., **94-4**, (131)

P

Paasivirta, J., **9-4**, (108) (109) (110) (112) (121); **15-4**, (109) (121) (125); **16-4**, (109) (119) (123); **17-4**, (109) (119) (126); **28-4**, (110) (111) (112) (121); **30-4**, (112) (115) (125); **37-4**, (112) (115) (125); **42-4**, (115); **43-4**, (115); **44-4**, (117); **54-4**, (123) (125); **70-4**, (131); **73-4**, (132) (133); **81-4**, (137); **84-4**, (133); **85-4**, (133) (135) (137) (138); **98-4**, (135); **101-4**, (137)
Partenheimer, W., **48-5**, (166)
Pascual, C., **83-2**, (49)
Pearce, H. L., **31-3**, (57) (65) (76); **12-4**, (109) (110); **18-4**, (110) (129)
Pearson, R., **23-1**, (3)
Perry, J. W., **44-2**, (21); **42-3**, (58) (67) (68) (75) (77) (78) (87) (89) (96)
Peters, J. A., **8-3**, (56) (58) (91); **48-3**, (60); **49-3**, (60); **65-3**, (67) (75) (77) (89) (91); **74-3**, (75); **76-3**, (75); **83-3**, (76) (79) (91) (95); **125-3**, (95); **12-4**, (109) (110); 15-5, (152) (154)
Petersen, M. R., Jr., **12-1**, (1)
Petrocine, D. V., **62-3**, (64)
Petrukhin, O. M., **26-3**, (57)
Phillips, T., II, **20-3**, (57); **21-3**, (57)
Pickering, M., **54-1**, (16)
Pijpers, F. W., **50-3**, (16)
Pirkle, W. H., **8-2**, (31) (49)
Pitt, I. G., **122-3**, (97)
Platzer, N., **27-5**, (152) (156) (157) (158) (170) (171) (172)
Porai-Koshits, M. A., **27-3**, (57)
Porter, G. B., **85-4**, (135) (137) (138)
Porter, R., **35-3**, (57)
Preston, R. K., **111-3**, (89)
Propeck, G. J., **39-3**, (58) (96); **42-3**, (58) (67) (68) (75) (77) (78) (87) (89) (96); **88-3**, (89) (96)
Proulx, T. W., **31-3**, (57) (64) (65) (67) (69) (70) (73) (74) (75) (76) (91) (92); **12-4**, (109) (110); **18-4**, (110) (129)
Pulkkinen, E., **26-4**, (110); **28-4**, (110) (111) (112) (121); **82-4**, (133)
Pyykkö, P., **55-3**, (61) (95); **4-4**, (107) (128)

R

Raber, D. J., **6-2**, (22); **41-2**, (21); **44-2**, (21); **45-2**, (21); **38-3**, (58) (61) (67) (93); **39-3**, (58) (96); **42-3**, (58) (67) (68) (75) (77) (78) (87) (89) (96); **43-3**, (58) (67) (68) (75) (77) (78); **44-3**, (58) (67) (68) (75) (77) (78); **52-3**, (60); **61-3**, (64) (65) (66) (67) (69) (70) (73) (74) (75) (76) (78) (89) (91) (92); **65-3**, (67) (75) (77) (89) (91); **66-3**, (67) (75) (77) (91) (94); **74-3**, (75); **75-3**, (75); **76-3**, (75); **77-3**, (75); **78-3**, (75); **83-3**, (76) (79) (91) (95); **88-3**, (89) (96); **93-3**, (89) (92); **124-3**, (95); **125-3**, (95)
Raber, N. K., **41-2**, (21); **42-3**, (58) (67) (68) (75) (77) (78) (87) (89) (96); **66-3**, (67) (75) (77) (91) (94); **78-3**, (75)
Rackham, D. M., **4-1**, (1); **57-2**, (24); **6-3**, (56) (58) (96); **71-3**, (75) (78) (89); **72-3**, (75) (78); **73-3**, (75) (78) (89); **3-5**, (151); **16-5**, (152) (159) (171); **20-5**, (152) (156) (159) (171)
Radionov, D. M., **23-5**, (152) (156) (166) (167) (169) (171) (172)
Rafaiski, A. J., **5-2**, (49)
Rajeswari, K., **113-3**, (89)
Ramage, R., **56-4**, (126) (128)
Randall, E. W., **15-2**, (21); **21-2**, (21); **25-2**, (21)
Ranganayakulu, K., **113-3**, (89)
Recca, A., **63-2**, (25); **72-4**, (131)
Reilley, C. N., **34-3**, (57)
Reuben, J., **7-3**, (56) (58); **14-3**, (57) (59) (75) (95); **53-3**, (61); **60-3**, (62) (73); **82-3**, (76) (77) (92); **1-4**, (107); **8-4**, (108); **57-4**, (128); **71-4**, (131); **4-5**, (151); **37-5**, (158)
Rewicki, D., **9-3**, (57); **91-3**
Rhine, W. E., **9-1**, (1)
Richardson, F. S., **102-3**, (87)
Richardson, M. F., **51-1**, (15); **27-2**, (21) (49); **15-3**, (57); **16-3**, (57); **32-3**, (57)
Riley, R., **53-5**, (169)
Rinaldi, P. L., **8-2**, (31) (49)
Ritchey, W. M., **49-5**, (166)
Robert, A. W., **72-2**, (49)
Roberts, D. W., **31-1**, (6); **32-1**, (6); **3-2**, (21) (49); **46-2**, (21); **72-2**, (49)

Roberts, J. D., **31-1**, (6); **32-1**, (6); **3-2**, (21) (49); **17-2**, (21) (24); **46-2**, (21); **72-2**, (49)
Robertson, R. E., **1-2**, (21); **5-4**, (108); **28-5**, (154)
Robinson, G., **53-4**, (122)
Rockefeller, H. A., **54-1**, (16); **87-3**
Romano, L. J., **110-3**, (89); **45-5**, (164)
Rondeau, R. E., **26-1**, (5); **49-1**, (15); **53-1**, (16)
Rose, A. J., **84-2**, (50)
Rossotti, F. J. C., **75-2**, (49)
Roth, K., **9-3**, (57); **91-3**
Rothstein, S. M., **27-2**, (6)
Russell, R. A., **122-3**, (97)

S

Sadlowski, J. E., **3-5**, (160) (162)
Saito, H., **55-1**, (16)
Sakkers, P. J. D., **24-2**, (21)
Sales, K. D., **15-2**, (21); **21-2**, (21); **25-2**, (21)
Sanders, J. K. M., **2-1**, (1) (5) (13); **21-1**, (2) (12); **29-1**, (5) (6); **35-1**, (6); **38-1**, (8); **82-2**, (49); **10-3**, (57) (62) (64); **63-3**, (64); **118-3**, (96)
Sands, D. E., **15-3**, (57); **16-3**, (57); **20-3**, (57); **21-3**, (57)
San Filippo, J., Jr., **110-3**, (89); **45-5** (164)
Sasaki, Y., **84-3**, (89); **89-3**, **96-3**
Schinco, F. P., **44-5**, (162)
Schneider, H. J., **60-2**, (24) (25)
Schuyl, P. J. W., **15-5**, (152) (154)
Schwalke, M. A., **6-2**, (22); **45-2**, (21); **42-3**, (58) (67) (68) (75) (77) (78) (87) (89) (96); **75-3**, (75); **77-3**, (75)
Seff, K., **34-1**, (6); **24-3**, (57) (58); **47-3**, (59)
Shapiro, B. L., **36-1**, (8) (9); **6-2**, (22); **35-2**, (21); **11-3**, (57) (75) (76); **31-3**, (57) (65) (76) (90); **77-3**, (75); **12-4**, (109) (110); **13-4**, (109); **18-4**, (110) (129)
Shapiro, M. J., **31-3**, (57) (65) (76) (90); **12-4**, (109) (110); **18-4**, (110) (129)
Sherry, A. D., **56-3**, (61) (95)
Shriver, D. F., **35-3**, (57)
Siddal, T. H., III, **34-2**, (21)
Sievers, R. E., **8-1**, (1); **9-1**, (1); **15-1**, (1) (13) (15); **26-1**, (5); **28-1**, (6); **49-1**, (15); **51-1**, (15); **53-1**, (16); **23-2**, (21); **16-3**, (57); **32-3**, (57); **32-3**, (57); **40-3**, (58); **3-5**, (151); **18-5**, (152) (156) (158) (165) (166) (167) (168) (169); **21-5**, (152) (156) (159) (160) (161) (162) (163)
Siltanen, E., **31-4**, (112) (115); **32-4**, (112) (115)
Silverstein, R. M., **42-5**, (160)
Simmie, J., **120-3**, (97)
Simmons, K. A., **8-2**, (31) (49)
Simpson, J., **85-3**
Sipe, J. P., III, **45-1**, (13); **29-2**, (21); **25-3**, (57) (58)
Smidt, J., **48-3**, (60)
Smith, G. V., 44-1, (13)
Smith, W. B., **17-5**, (152) (156) (158)
Sneen, R. A., **46-4**, (119)
Snyder, G. H., **20-2**, (21)
Sokalski, W. A., **48-2**, (22)
Souza, S. M., **54-5**, (169)
Springer, C. S., Jr., **28-1**, (6); **54-1**, (16); **32-2**, (57); **87-3**
Staley, R. H., **104-3**, (89)
Stalling, W. C., **111-3**, (89)
Staniforth, M. L., **21-2**, (21); **25-2**, (21)
Stefaniak, L., **39-1**, (10) (12)
Stemple, N. R., **23-3**, (57)
Stilbs, P., **47-4**, (119)
Stothers, J. B., **104-4**, (140)
Strömberg, S., **68-4**, (131)
Suchánek, M., **37-3**, (57)
Sugimura, T. S., **55-1**, (16)
Sullivan, G. R., **10-1**, (1); **36-1**, (8) (9); **35-2**, (21); **53-2**, (22) (49); **52-5**, (167)
Suontamo, R., **85-4**, (133) (134) (137) (138)
Surma-Aho, K., **85-4**, (133) (134) (137) (138)
Sutton, P., **38-3**, (58) (61) (67) (93); **42-3**, (58) (67) (68) (75) (77) (87) (89) (96)
Sweigert, D. A., **33-1**, (6); **47-2**, (22)
Sykes, B. D., **20-2**, (21)

T

Taagepera, M., **104-3**, (89)
Taft, R. W., **104-3**, (89)
Tai, J. C., **45-4**, (119)
Takahashi, S., **79-3**, (75)
Takagi, T., **84-3**, (89)
Talvitie, A., **14-4**, (109) (119) (121) (131); **66-4**, (131); **67-4**, (131); **68-4**, (131); **69-4**, (131); **70-4**, (131)
Tamburin, H. J., **36-2**, (21) (25)
Tanner, L., **25-5** (152) (156) (157)
Tanny, S. R., **54-1**, (16); **87-3**
Tasumi, M., **79-3**, (75); **80-3**, (75)
Tatsuo, M., **3-5**, (151)
Taylor, R. C., **50-5**, (162); **51-5**, (166)
Templeton, D. H., **18-3**, (57)
Thakkar, A. L., **76-2**, (49)

Thomas, J. R., **41-4**, (115)
Thompson, H. B., **10-2**, (22); **13-2**, (21)
Thornton, J. M., **56-2**, (23)
Toivonen, N. J., **22-4**, (110); **23-4**, (110); **31-4**, (112) (115); **32-4**, (112) (115); **33-4**, (112) (115); **34-4**, (112) (115)
Tori, K., **41-1**, (12) (13)
Towns, R. L. R., **12-4**, (109) (110)
Trifan, D. S., **74-4**, (133)
Trueblood, K. N., **35-5**, (157)
Tucker, J. N., **3-5**, (151)
Tuovinen, M., **31-4**, (112) (115)
Tupper, D. E., **16-5**, (152) (159) (171)

U

Uchio, Y., **50-4**, (119)
Uebel, J. J., **19-2**, (21); **22-2** (21) (49)

V

Van Bekkum, H., **49-3**, (60); **15-5**, (152) (154)
Van Bruijnsvoort, A., **109-3**, (89) (96)
Van der Kerk, G. J. M., **39-5**, (159) (160) (170)
Van Der Puy, M., **64-4**, (129) (130)
Van Dongen, J. P. C. M., **43-5**, (162)
Van Emster, K., **102-4**, (137)
Van Willigen, H., **20-1**, (2)
Vaughan, W. R., **104-4**, (140)
Vecchio, G., **109-3**, (89)
Vodicka, L., **58-2**, (24); **62-4**, (129); **106-3**, (89)
Volka, K., **37-3**, (57)
Von Ammon, R., **7-1**, (1) (6) (10)
Von R. Schleyer, P., **83-4**, (133)
Von Rudloff, E., **48-4**, (119)
Von Schantz, M., **49-4**, (119)
Von Sprecher, G., **96-4**, (135)

W

Wagner, W. F., **46-1**, (14); **15-3**, (57); **20-3**, (57); **21-3**, (57)
Wahl, G. H., Jr., 12-1, (1)
Waigh, R. D., **68-2**, (25)
Waldner, A., **96-4**, (135); **1004-**, (135)
Walters, D. B., **51-5**, (166)
Warner, P., **46-1**, (115)
Warrener, R. N., **122-3**, (97)
Wasson, J. R., **16-1**, (1)
Waterman, P. G., **68-2**, (25)
Watkins, E. D., II, **20-3**, (57)
Watson, W. H., **23-3**, (57); **34-5**, (154)
Watt, P. H., **36-2**, (21) (25)
Weber, M., **68-4**, (131)
Weigand, E. F., **60-2**, (24) (25)
Weiss, F. T., **40-5**, (160) (170)
Welti, D. H., **61-2**, (24)
Wenkert, E., **25-1**, (5) (6); **64-3**, (64); **10-4**, (108) (115) (137); **61-4**, (129)
Wenzel, T. J., **13-5**, (152) (159) (162); **14-5**, (154) (155) (156) (157) (158) (159) (164) (165) (169) (170) (172); **18-5**, (152) (156) (158) (165) (166) (167) (168) (169); **21-5**, (152) (156) (159) (160) (161) (162) (163); **24-5**, (152) (155) (156) (159) (166) (167) (169); **36-5**, (157) (171); **55-5**, (172)
Werbelow, L. G., **81-2**, (49); **68-3**, (69) (71); **11-4**, (109) (129)
Werstiuk, N. H., **90-4**, (133) (135)
West, R., **53-5**, (169)
White, A. M., **76-4**, (133); **77-4**, (133); **99-4**, (135)
Whittaker, E. T., **53-4**, (122)
Widen, K.-G., **17-4**, (109) (119) (126); **54-4**, (123) (125); **70-4**, (131)
Wijekoon, W. M. D., **69-4**, (131)
Willcott, R. M., III, **40-1**, (10); **9-2**, (31); **18-2**, (21) (22) (24) (49); **43-2**, (21); **73-2**, (49); **123-3**, (97); **20-4**, (110); **55-4**, (126)
Willcox, C. F., Jr., **51-4**, (121)
Williams, D. H., **11-1**, (1) (4) (12); **21-1**, (2) (12); **29-1**, (5) (6); **30-1**, (6); **38-1**, (18); **1-2**, (21); **65-2**, (25) (49); **82-2**, (49); **10-3**, (57) (62) (64); **63-3**, (64); **118-3**, (96)
Williams, R. J. P., **33-1**, (6) (12); **42-1**, (12); **43-1**, (12); **42-2**, (21); **56-2**, (23); **75-2**, (49); **23-3**, (57)
Williamson, K. L., **60-4**, (129)
Wilson, C. L., **103-4**, (137)
Wing, R. M., **19-2**, (21); **22-2**, (21) (49)
Winstein, S., **74-4**, (133)
Witanowski, M., **39-1**, (10) (12)
Wolf, J. F., **104-3**, (89)
Wolkowski, Z. W., **37-1**, (18); **39-1**, (10) (12)
Wright, L. L., **25-5**, (152) (156) (157)
Wyatt, M., **22-1**, (3) (15); **30-3**, (57) (59); **45-3**, (59)

X

Xavier, A. V., **42-1**, (12); **75-2**, (49)

Y

Yang, P. P., **56-3**, (61) (95)
Yoshimura, Y., **41-1**, (12) (13)
Youngs, D. S., **47-1**, (14)
Yurchenko, A. G., **23-5**, (152) (156) (166) (167) (169) (171) (172)

Z

Zalkin, A., **18-3**, (57)
Zimmerman, D., **92-3**
Zvolinskii, V. P., **74-2**, (49)

Subject Index

A

Acetophenones, 89
Acid-catalyzed racemization of camphene, 140
Alcohols, isomeric, 115–117
  dehydrocamphor, 115
  Grignard reaction, 115
  hydroxyls in, 115
  2-methylisoborneol, 115
  proton spectra, 115
Alcohols, position of lanthanum ion with respect to, 24
Alcohols, primary, diastereotopic groups
  aliphatic acyclic alcohols, 128
  aliphatic protons, 128
  conformers, 128
  cyclic methanols, 126
  (3)-cyclopentenyl-1-methanol, 126, 127
  diastereotopy, 128
  3-endo-methyl-5-norbornene-2-exo-methanol, 126, 127
  and europium on α protons, 128
  3-exo-methyl-5-norbornene-2-endo-methanol, 126, 127
  and LANTA iteration, 128
  methylene-group protons, 126
  (—)-myrtenol, 126, 127
  5-norbornene-2-*endo*-methanol, 126, 127
  pseudocontact effect, calculation about, 128
  stable conformations, 126
Alcohols, tertiary, 1:1 $Eu(dpm)_3$ complexes, average parameters, 125
Aldehydes, and lanthanum ion, 25–26
Alkanes, polyhalogenated, 89
Alkoxyl compounds, and lanthanum ion, 24
Alkyl fluorides, 89
Alkyl groups, tertiary, and lanthanum ion, 24
Amines, and lanthanum ion, 24, 89
Angular term, 8, 128
Anilines, 89
Aromatics, and binuclear reagents
  anthracenes, 162
  binding to Ag(I), 159
  ^{13}C NMR spectra, 162, 164
  4-bond coupling, 160
  fused ring system without substituent group, 162
  long-range coupling, absence of, 159
  9-methylanthracene, 162, 163
  methylbenzene derivatives, 160
  naphthalenes, 162
  polycyclic compounds, 162
  and $Pr(fod)_3$-Ag(fod), 159
  proton couplings, 159
  pyrenes, 162
  resonances of aromatic products, 159–160
  steric effects, 159
  p-terphenyl, 162
  toluene, NMR spectra, 159
  unleaded gas, analysis, 160, 161
  xylenes, 160, 161
Aromatics, substituent effects in, 89
Aryl fluoride, 89
Association constants
  alcohols, 82–83
  aldehydes, 81
  amides, 82
  amines, 84–85
  amine oxides, 84
  carboxylic acid derivatives, 82
  chelating (difunctional) compounds, 86
  cyclohexanols, substituted, 87
  esters, 82
  ethers, 83
  ketones, 81
  nitriles, 85–86
  phosphine oxides, 84
  pyridine, 84
  vs. solvents, 88
  sulfoxides, 84

B

Basicity of substituent, 3
Binuclear lanthanide (III)-silver (I) NMR shift reagents: *See also specific type of reagent*
  achiral shift reagents, 155
  and Ag(fod), 152, 155
  and Ag(tfa), 152, 155
  atom size as factor, 155

attraction mechanism, problem of, 154
chiral analogs, 152
β-lanthanide chelates, ineffective, 156
diketonate, 1:1 complex in solution, 154
discovery of, 151–152
donor, holding of, 152
geometrical change vs. ligand, 155–156
1-hexene, 155
hydrogen ligands, 156
improved spectral dispersion, applications, 155
KBF_4, 154
lanthanide cesium complexes, 154
ligand choice vs. magnitude of shifts, 155
ligands used, 153
$Lu(fod)_3$, 154
methine carbons, 154
nickel(II)-silver(I) diketonates, 154
olefin, association with silver, 155
parent ketone, 153
platinum complexes, 154
and $Pr(fod)_3$, 152, 154
pseudocontact shift equation, use, 155
silver as bridge, 152
silver ligands, 156
and soft Lewis bases, 152
tetrakis chelate anions, 154
toluene NMR spectra, 154
$Yb(fod)_3$, 155
Binuclear reagents, solvents for
bad choices, 172
good choices, 172
Binuclear relaxation agent, 137–138
Bonding constant vs. basicity, correlation, 87–88
Borneol, 110
Bound shifts, direct observation of
DMSO, substitution into, 59
$Eu(dpm)_3$, 59
$Eu(fod)_3$, 59
slow exchange, 59
and strong Lewis bases, 59
Bound shifts, extraction from solution data, *See* Bound shifts; Equimolar method; Gradient method; Reciprocal method; Two-step method
2-Butylamine, LIS with ^{13}C, 12

C

Camphene, 137, 139, 158
cations of, 137
Carbon-13, bound shifts of, 7, 94–95
carbon spectra, 95
problems with, 94–95
proton spectra, 94, 95
and two-step method, 95
Carbonyl compound complex coordinates, 26
Carboxylic group:
diagram, 24
lanthanum ion, position of, 23–24
Cartesian coordinates, 22–23
CCl_4 solution, LIS values, 4
Cesium, 154
Chiral substratent and binuclear reagents
Ag(facam), 168, 169
Ag(fod), 168, 169
Ag(hfbc), 168, 169
Ag(hfth), 169, 170
and $AgNO_3$, 168
Ag(tfa), 169
Ag(tta), 169
aromatics, studies on, 167–168
association constant vs. enantiomer and reagent, 167
camphene, results with, 168
and chiral diketonates, 166
complex geometry, enantiomer and reagent, 167
cyclohexane derivative, olefinic protons, 167
downfield reagent choice, 170
enantiomer resolution, 167, 168
and $Eu(facam)_3$ + silver trifluoroacetate, 167
geometries of complexes vs. shift magnitude, 168
and H(facam), 167
and H(hfbc), 167
light, silver diketonate sensitivity to, 169
olefins, 167–168
α-pinene results, 168
$Pr(facam)_3$, 168, 170
$Pr(hfbc)_3$, 168, 170
proton resonances, 167
relative magnitude of shifts, sequence, 168–169
silver complexes, 168
and silver species, enantiomeric resolution vs., 167
spectra to record, points to remember, 167
stability order, silver reagents, 169
upfield reagent, choice, 169–170
$Yb(facam)_3$, 169
ytterbium, complex of, 167

Cholesterol, LSR complexes and shifts of, 12
Complex:
    atomic structure, 6
    formation shifts, 7, 12
    magnetic axis in, 6
Concentration ratio, in equilibrium method, 92
Contact shift:
    formula, 7
    mechanism, 7–8
    vs. pseudocontact shifts, 7, 11
Cyclic ether, lanthanum ion's place in, 26
3-Cyclopentenyl-1-methanol, 126

D

Deuterium label stereochemistry
    and active hydrogen in reactant, 132
    camphene, reaction, scheme, 139
    corner, protonation, 137
    dehydronorcamphor, 132
    deuterated isoborneol, 138
    deuteration of double bond of norbornenes, 132
    deuteration, multiple, 138
    dideuteroformic acid, 137, 138, 139
    exo attack, 140
    exo-norbornyl derivatives, 133
    formates, 137
    and formic acid, 133, 137
    gegenions, 135
    graded σ participations, 135
    isobornyl chloride, 137
    isobornyl formate, 137
    and LISNMR method, 135
    Nametkin rearrangement, 140
    2-methylnorbornyl ion structure, 135
    1-methylnortricyclene, 137
    norbornan-2-*exo*-ols, 133
    norbornane-2-*exo*-formate, 133
    norbornane skeleton signals, 137
    norbornene deuteration, 135
    norbornene, reactions of with acids, 133, 135
    2-norbornyl carbenium ion, 133
    2-norbornyl ion, 133
    nortricyclene, 133
    proton NMR spectra, 132
    saponified reaction products, 134, 137, 138
    σ-bond participation structures, norbornyl cations, 136
    σ-bridging, 133, 135
    and signals, deuterlum distribution, 133, 135
    stereospecific rearrangements, tertiary cations, 139–140
    3*x*-7*S* distribution, 135
    tricyclene, reaction scheme, 137, 138
    Wagner-Meerwein rearrangement, 140
Diamagnetic ions, triply-charged, 2
Diastereotopy, 126–129
Dipolar term, 107–108
DMSO, $Eu(fod)_3$ complex with, 15, 59
Donor with 2 donor sites, 28
Dy^{+3} ion, 2
Dysprosium reagents, 166

E

Electrons:
    donation of, 89
    spin relaxation times, 2
    withdrawals of, 89
Enantiomers, resolution of, 167–168
Epoxides, lanthanum ion location in, 26
Equilibria, LSR, solution of, 4
Equimolar method, bound shifts
    advantages, 91–92
    diagram, 73
    disadvantages, 92
    formula, 73
    hypothetical compounds, results, 73–75
    recommendations, 80
    shift reagent, bonding to, 74–75
    slope, 73
    variables, optimization of, 76
Europium
    advantages of, 4–5
    downfield protons, 5
    and methyl oleanate, 5
    and praseodymium, 5
    triply-charged ion, 2
$Eu(dfd)_3$, 14
$Eu(dfhd)_3$, 15
$Eu(dfhd)_3 \cdot (H_2O)_2$, 15
$Eu(dfhd)_3 \cdot (H_2O)_{0.5}$, 15
$Eu(dpm)_3$
    complexes, 15
    diagram, 1
    dipyridine complex, 1
    nature of, 1
    shifting power, 14
    solubility, 5
$Eu(dpm)_3(py)_2$, 6
$Eu(fhd)_3$, 14

$Eu(fod)_3$
reagents for, 90–91
shifting power, 14
solubility of, 5
structure, 1
$Eu(pfd)_3$, 14
$Eu(tfn)_3$, 14

F

Fermi contact distribution, 10–11
Fluorinated ligands, 15
Fluorine, single atoms, LISs of, 89
Fused ring polycyclic aromatics, 162

G

Gadolinium, 12
$Gd(fod)_3$, 110
Geminal protons, diastereotopic, 128
Gradient method, and bound shifts
approximations, 65
error range, 66–67
and $Eu(fod)_3$, 67
experimental use, 64–65
and (fod) derivatives, 65
free substrate, 65, 66
graphic plot, discussion, 64–65
hypothetical compounds, results with, 66–69
1:1 complex formula, 65
and association constant, 65
1:2 complexes, 65
other factors in results, 66
problems with, 69
substrate vs. adduct ratio, 69
and $Yb(dpm)_3$, 66

H

H(facam), 167
H(hfbc), 167
H(tfn), 15
Halogenated compounds, and binuclear reagents
bromopentane, 165
t-butyl resonance, 166
and C=C bond, 165
chlorobutane, 165
chlorocyclohexane, 165
chloropentane, 165
compounds unsuited for, 164
$Dy(dfhd)_3$-Ag(fod) reagent, 166
$Dy(fod)_3$-Ag(fod), 164
halide substrate-silver, association constant, 164
halobutanes, 165
halogen atom sizes, 164–165
iodobenzene, 164
iodohexane, 165
precipitates, 164
and silver, 164
tris-chelates, 164
Halogen substituents, coordination to, 89
1-Heptanol, proton NMR spectra, 3
Heterocyclic bases, tertiary, lanthanum ion place in, 24–25
Ho^{+3}, 2
Hydrogen LIS shifts, 7
Hydrogen NMR shifts, formula, 6

I

Induced shifts, simulation of
borneol isomers, 21
isomerism, 22
protons, 21
Isoborneol, 110, 137, 138
Iterative computer program, use for finding lanthanum position, 119

K

Ketones, lanthanum ion place in, 25–26

L

L + S equilibria, 57
Lactams, lanthanum ion's place in, 25–26
Lactones, lanthanum ion's place in, 25–26
LANTA program, 145–150
Lanthanide-induced shifts (LISs), 2–4
and substrates in CCl_4 for solutions, values found, 4
Lanthanide-induced shifts, bound shifts extraction from data
acetoxy methyl, hydrogens of, 61–62
1-adamantyl acetate, 61–62
algebra about, 63
approximation method, 64
complexes, 61
contribution formula, complex, 63
formula for extraction, 63
methylene hydrogens, 62
numerical values for nuclei, 61

quinoline, 62
shift reagent equilibria formulas, 63
2-step method, 63–64
Lanthanide-induced shifts, with ^{13}C, 12
Lanthanide-induced shifts, electron pairs in rigid structures, 110–119
Lanthanide-induced shift effects, theory of
components, 107
and magnetic susceptibility anisotropy, 108
pseudocontact equation, 108
separate proton signals from skeleton, 108
stereostructures, deduction of, 108
Lanthanide-induced shifts, experimental remarks
and ^{13}C NMR, 110
constant-S_0 incremental dilution method, 109
dryness, need for, 4, 109
and $Eu(dpm)_3$, 110
gas chromatography, 109
$Gd(fod)_3$, 110
and ^1H NMR, 109–110
plots, 109
proton, chemical shift dependence on LSR linearity, 109
substrate-reagent solutions, 108–109
$Yb(dpm)_3$, 110
Lanthanide shift reagents. *See also specific reagent*
analysis, common method, 2–3
complex, Cartesian coordinates for, 27
solubility of, 2
spherical coordinates for, 27
usual proton resonances, 5–6
Lanthanides
contact, 10, 11
methylene groups, shifts in cyclohexane, 10
pseudocontact, 10, 11
Lanthanum ion
description, 2
position in complex, 23. *See also specific chemical*
Lanthanum *tris*-chelates, 151
Lanthanum *tris*-β-diketonates, hard Lewis acid, 151
Lewis acids
and bound shifts, 90–91
metal cations, 2
Lewis bases
and polyfunctional substances, 170–171
soft, 58, 148, 170–171
strong, 58, 59
Lewis basic functional groups, 2
Line broadening
and bound shifts, 90
by metal ions, 10, 11
LIS formula, 108
LIS signature, 9, 10
$Ln(dpm)_3$, 6
$Ln(fod)_3$, studies of, 16
Low-temperature spectra of stoichiometric complexes
chromium, 60
coalescence temperatures, 61
1,2-dimethoxyethane, 60
hexamethylphosphoramide, 59
3-methylpyridine, 59
4-methylpyridine, 59
polyethyleneglycol ethers, 60
pyridine, 59
quinuclidine, 60
slow exchange, problems with observation, 60–61
tetramethylurea, 59
trans-metal acetylacetonates, 60
triethylamine, 59
LS + S equilibria, 57
LSR association, 3
LSR solubility, 15
LSR + S equilibria, 3, 4
Lutetium ion, 2

M

Magnetic anisotropy, reasons for, 2
Magnetic axis, angle dependence of, 6, 9
McConnell-Robertson equation, 6, 21, 108
Metal-substrate equilibrium, 3
Methyl benzoates, substituted, 89
Methyl oleanate, 5
9-Methylanthracene, 163
Methylbenzene derivatives, 160
Methylbicyclo[2.2.2] octenols, stereoisomers, 126
2-Methylenenorbornane, 136
2-Methylisoborneol, 115
1-Methylnortricyclene, 136, 140
p-Methoxy-*p*′-methylazoxybenzene, 13
Modified gradient method, bound shifts extraction
advantages of, 93, 94
and bound shift inaccuracy, 93

lanthanide/substrate ratios, 93–94
problems with, 92–93
sample preparation, 93
scaling errors, 94
Monofluoroalkyl groups, 89
Multifunctional substrates, 170–171

N

Naphthyl protons, downfield shifts, 8–9
compound structure, 8
results, 9
Naphthyl protons, upfield shifts, 9
Newton-Raphson iteration method, 122
Nitriles, methyl substitution, 90
Nitrocompounds, 15
NMR solvents, 5
NMR, tube, calibration of, 92
Nojigiku alcohol
apoisofenchol structure, 123
and *Chrysanthemum* species, 119
and fenchols, comparison with, 121
fitting of results, 123
α-isofenchol structure, 122
and LANTA program, 121–122
linear equations, 122–123
proton NMR spectra, 120
relative LIS values, 121
structure 120, 121
derivation of, 121
Nonlinear LIS plots, reasons for, 4
Norbornan-2-*exo*-ols, nondeuterated, 111, 133, 134
Norbornane, 110
Norbornane-2-*exo*-formate, 133
2-Norbornanols, 110–114
Norbornene, 115, 133, 134
5-Norbornene-2-*endo*-methanol, 127
2-Norbornenols, 112
2-Norbornyl cation, nonclassical, 133–135
2-Norbornyl ion, structure and chemistry of. *See also* Deuterium label stereochemistry
and α-protons, 110
coupled protons, 111
coupling constants, 112
dehydronorborneols, 112
endo-2-norbornanol
proton NMR spectra, 112
structure, 112
exo-2-norbornanol,
proton NMR spectra, 111
structure, 111
fenchols, 110
norbornane, 110
and norcamphor, 115
proton NMR spectra, problems with, 110–111
terpenes, 110
5,6-unsaturated 2-norbornenols, 112
Norcamphor-5-*endo*-6-*endo*-d_2, 133
Nortricyclanols, *cis-trans* stereoisomerism in structures
bridge methylene protons, 115
pericyclobornanol, 118
proton NMR spectra, 118
proton signals from, 115, 117, 118
Nortricyclene, 133, 135
edge-protonated ion, 135
Nuclei, magnetic activity of, shifts of, 12
Nuclei in substrate, sequence of simulation, 10

O

Olefins, and binuclear reagents
^{13}C NMR study, 156–159
cis-isomers, 156, 157, 158
contact shifts, 158
$Gd(fod)_3$-Ag(fod), 157
and ^1H NMR spectral shift, 156
hexene isomers, 157, 158
nonene isomers, 157
and 2-octene isomer mixture, spectra, 156–157
olefins isomer mix, 157
pseudocontact mechanism, 158
reasons for, 157
relaxation reagent, 158
shift vs. metal of reagent, 158
shift, concentration vs. relative magnitude of shift, 156, 157
trans-isomers, 156, 157, 158
Oxygen, and lanthanum ion, 28

P

Paramagnetic ions, 2
Pearson acids and bases, hard and soft, concept, 3–4
Pentane as solvent
resonance overlap, 172
shift increase by, 172
Pericyclobornanol, 117

Phosphines
and binuclear reagents, 166
lanthanide *tris*-β-diketonates, 166
NMR spectra, ^{31}P, 166
triphenylphosphine spectra, 166
Polarizable substances, nuclei in, shifts of, 10
Polyethylene glycol ethers, 60
Polyfunctional compounds, bound shifts of
aromatic rings, 96
blocking groups, and conversion to monofunctionality, 96
bonding strength, and chelation, 97
t-butyldimethylsilyl ether, 96
chelate rings, 96
cooperative bonding between functional groups, 96–97
8-coordinate lanthanide, 97
fluorenones, 96
ketone group, 96
mathematical analysis, 95–96
oxygen heterocyclics, 97
steroid derivatives, 96
trifluoroacetates, 96
Polyfunctional substrates and binuclear reagents
Ag(I) bonding, 170
benzothiophenenes, sulfur in, 171
benzoxazoles, 171
bitter substances in mushrooms, analysis of, 171
hard Lewis, bonding to, 171
hard and soft Lewis donor groups in compound, 170
heteroaryl system without lanthanide reagent shifts, 171
indene, 170
linalool, analysis, 171–172
4-methylstyrene, 170
nonane derivative, ^{1}H NMR spectra, 171, 172
olefin bonding, 170
π-electrons, 170
reagent, bonding site of, 170–171
soft Lewis bases, 170–171
bonding to, 170–171
styrene, 170
Praseodymium
ion, 2, 4, 5
reagents and bound shift, 90
$Pr(dfhd)_3$, 15
$Pr(facam)_3$, 168, 170
$Pr(fod)_3$ complex, 15
$Pr(hfbc)_3$, 168, 170
Protons
affinities, 87, 89
aromatics, coupling in, 159
^{13}C spectra, 94, 95
chiral substances, resonances, 167
and deuterium label, 132
and europium, 5
irradiation resonance, 113
and LSRs, 5–6
naphthyl, 9
NMR shifts, formula, 6
nonequivalence, enhancement, 2
2-norbornyl ion, 110–111
nortricyclanols, 115–119
olefins, 156
skeleton signals, 108
Pseudocontact mechanism
check on, 12
lanthanides, 10, 11
LSR + S complex, 6, 7
Pseudocontact shift, early work on, 8
Pyridine, lanthanum ion's place in, 24–25

R

Reagent discrepancies and bound shifts, 91
Reciprocal method and bound shift extraction from data
association constants, 72
bound shift slope, 70
complexation percentages, 72
complexes, stoichiometry of, 70
error range, 72
formula, 69
hypothetical compounds, results with
adduct rates, 71–72
plot, 71
substrate, 71, 72
low $[L_0/S_0]$ ratios, 69
recommendations, 80
Ring conformations
4-*t*-butylcyclohexanols, 129, 131
4-*t*-butylcyclohexanone, 131
cadinene derivative, 131
α-cadinol, 131
T-cadinol, 131
2-*cis*-methyl-4-*t*-butylcyclohexanone, 131
cubenol, 130, 131
cyclohexane, 129
cyclohexanols, 129
cyclohexanones, 129–130
equilibrium constant for, 129

△-Eu values for, 130
deuterated, 129
epicubenol, 130, 131
hinokiic acid, 131
and hydrogen bonding vs. equatorial-axial ratio, 129
4α-hydroxy-10(1)-cadinene, 131
ketones, cyclic, 131
T-muurolol, 131
sesquiterpenoid alcohols, 129–130
shift effect vs. equatorial-axial equilibrium, 129
torreyol, 131
valeranone, 131

S

Shift, origins of
contact interaction, 20
delocalization of free electron density, 7
dipolar interaction, 20–21
magnetic axis, 21
McConnell-Robertson equation, 21
Shift reagent equilibria, stoichiometry of complexes
adamantane derivatives, 58
complex formation, 1:1, 57, 58
dimerization, 57–58
$Eu(dpm)_3$, 58
$Eu(fod)_3$, 58
experimental work, 57
lanthanide shift reagent, interaction with organic substrate, 57
self-association of shift reagent, 57
solution, complexes in, 57
species mix, 56
strong Lewis bases, 58
time-average spectra, 56–57
weak Lewis bases, 58
Sigma (σ) bridging, 131, 134
Signal assignment, 122–126
Silver
as bridge, 152
in CS_2, 172
diketonate, 169
(facam), 168, 169
(fod), 168, 169
ligands, 156
(I)heptafluorobutyrate, 151
(hfbc), 168, 169
(hfth), 169, 170
nitrate, 168
terpenes, 158
(tfa), 169
(tta), 169
SIMULATION program structure and use
ADJUST subroutine, 31
APL, language used, 49–50
Cartesian coordinates, 27
diagram, 29
DIAX (1-HC) simulation, 31–33
DIAX (1-HC) with lanthanide positions simulation, 37–39
DIEQ (HC-I) simulation, 33–35
DIEQ (HC-1) with lanthanide positions simulation, 39–42
dihydropyran derivative, structure and example, 30
dipolar shifts, calculation of, 26–27
equilibria, differences between DIEQ (HC-1) and DIAX (1-HC), 35–37
example, diagram, 31
lanthanide motion in cube, induced shift range, 42–43
methyl groups, 31
and mobile protons, 31
new reference frame definition, method, 28–29
programs, alternate
CHMSHIFT, 49
2-conformer ratio, 49
grid shift scheme programs, 49
LANTHAMATIC, 49
MAXI, 49
Newton's method for lanthanide optimum position, 49
OPLIS, 49
PDIGM, 49
programs, list
ADJUST, 45
PGM1, 44
PGM2, 44–45
SIMN, 43–44
subroutines
CALLDATA, 45–46
CALLLANT, 46
CONVERT, 46–47
FACTOR, 47
MAXMIN, 48
OFFSIDE, 48
PLANE, 47
POSITION, 47
ROTATE, 48
triplets of numbers, 28, 29

Solvents
effects, and bound shifts, 91
preferred, 5
Staggered conformers, 128
Steric effects, 89
Stoichiometry of complexes, 56–58. *See also* Shift reagent equilibria
Symbols and abbreviations, list, 56
Symmetry, planes of, and C—X bond, 24

T

Terpenes, and binuclear reagents
camphene, 158
δ-3-carene, 158
d-limonene, 158
α-pinene, 158
β-pinene, 158
ring, steric effects on, 158
silver, bonding location of, 158
Tertiary cations, 139
Tertiary 2-methyl analogs
endo version, 114
exo version, 113
and hydroxyls, 115
proton irradiation, 113, 114
Time-average spectra, 56–57
Transition metal ions, line broadening of, 2
Tricyclene, 139
Trifluoroacetate fluorines, 89
Trifluoromethyl group, 89
Two-step method, bound shift extraction from data
association constant, size of, 75
too large constant, compounds with, 75
carbonyl compounds, 75
complex ratios vs. calculated bound shifts, 75
complexation range, 77, 79
equilibrium parameters, 76
vs. equimolar method, 79
$Eu(fod)_3$ studies, 75
example, diagram, 77
failure, reasons for, 79
fitting procedure, data for, 76
fitting, results, 79
hypothetical compounds, 77–78
nitriles, 75
recommendations, 80
solvent choice, 75
small number of data points, dealing with, 79–80
two-step procedure, 76–77

V

7-Valeranol, sesquiterpenoid, structure of
angle term, 123.....
computer iteration results, 125
correlation coefficients, 124, 125
europium-oxygen distances, 126
isopropenyl group protons, 123
methyl groups, 123
proton NMR spectra, 123, 124
diagram, 124
signal motion in field, 123
species, 123
structures, 123
and terpinen-4-ol, 123, 124
trans-decalin skeleton, 123

W

Water, problems with, 4

X

Xylene, 161

Y

Yb^{3+}, 2, 11
$Yb(dfhd)_3$, 15
$Yb(dpm)_3$, 110
Yb(facam), 169
$Yb(fod)_3$, 89, 154
Ytterbium reagents, 12